S0-BIB-898

THE OFFICE TELEPHONE

A User's Guide

THE OFFICE TELEPHONE

A User's Guide

Patricia A. Garner

Director of Education
United College of Business

Prentice-Hall, Inc., Englewood Cliffs, New Jersey 07632

Library of Congress Cataloging in Publication Data

Garner, Patricia A.
The office telephone.

Bibliography: p.
Includes index.
1. Telephone in business. 2. Telephone etiquette.
I. Title.
HE8735.G37 1984 651.7'3 83-13797
ISBN 0-13-631481-3

Editorial/production supervision and
interior design: Pamela Wilder
Cover design: George Cornell
Manufacturing buyer: Ed O'Dougherty

Printed in the United States of America

10 9 8 7 6 5 4 3 2 1

ISBN 0-13-631481-3

Prentice-Hall International, Inc., *London*
Prentice-Hall of Australia Pty. Limited, *Sydney*
Editora Prentice-Hall do Brasil, Ltda., *Rio de Janeiro*
Prentice-Hall Canada Inc., *Toronto*
Prentice-Hall of India Private Limited, *New Delhi*
Prentice-Hall of Japan, Inc., *Tokyo*
Prentice-Hall of Southeast Asia Pte. Ltd., *Singapore*
Whitehall Books Limited, *Wellington, New Zealand*

for **Mark**

Contents

8 Communication Strategies for Business 145

9 Costs and Controls 165

10 Technology 181

11 The Future 204

Appendix

A Resources 214

B Glossary 219

Index 233

Preface

The need for a book on how to use the telephone is like the weather (pardon the well-worn cliche); everyone talks about it but no one ever does anything about it. Volumes upon volumes have been written on the invention and development of the telephone, how it works, and what it can do, but relatively little has been written on how to use it. With the advent of competition in the telephone industry, the need is becoming even greater.

As I began to talk with others about the opportunity I had to do something about this lack, I found two universal feelings: Managers feel that telephone costs are too high, and users feel that other people are very impolite on the telephone. In fact, nearly everyone I spoke with—especially on a casual basis—said that there were a lot of people in their offices who need such a book. I found it increasingly humorous that so many of us see shortcomings in how others use the telephone but few of us admit to our own. I am as guilty as anyone.

These, however, are merely the stresses of everyday human relations and business, and they will be with us forever. All of us must continually strive to improve our communications skills and to be cost-conscious at work. Most certainly, an important goal of this book is to provide information on how to maintain good human relations and polite communications in business, both on and off the telephone. Chapter 9 is intended to bring usage costs into good perspective.

The far more important goal of the book, however, is to prepare office workers for the expanding role of telecommunications in their daily lives and tasks and to point up the importance of the office, whose very product is information, in the coming of the Information Age.

The merging technologies of data processing, word processing, micrographics, and reprographics are being linked by telecommunications, thus making the telephone much more than a simple tool whose proper use is obvious. Ken Asten, President of Ken Asten & Associates, Inc., and Academic Consultant for the MBA Telecommunications Program at Golden Gate University in Los Angeles, summed it up admirably when he said:

> There was a time when a business that needed telephone service simply called the telephone company and placed an order. There was also a time when the computer sales representative knew much more than the customer's data processing staff.
>
> Those days are rapidly disappearing. So is the day when a telephone company representative can promptly and objectively tell you which equipment will best meet your needs. And so is the day when the computer vendor can objectively tell you how to properly interface his equipment to communications lines and terminal equipment furnished by other suppliers.
>
> In the past 12 years there have been a tremendous number of developments in the telecommunications and teleprocessing fields that give greater choice for the user, but only if he is aware of the range of services that is available.

Rex Licklider, President of the Los Angeles-based Com Systems, Inc., framed the broader picture when he said:

> Today's business phone is the doorway to the transmission of anything from a sales proposal to an engineering drawing to a computer data file, instantly, through time and space.
>
> The telephone itself has become an increasingly "intelligent" device, capable of performing time- and labor-saving functions unheard of just a few years ago, as well as delivering impressive convenience features. As a result, explosive growth is the future of our telecommunications industry."

Acknowledgments

When I finally got off the fence and made an irrevocable commitment to this project, I was certain there would be moments when I would sorely regret that I submitted myself to pressures and long hours of work. Much to my delight, I was totally wrong. The entire project turned out to be completely enjoyable, primarily due to the excitement and sense of progress that permeates the telecommunications industry. The pressures push upward rather than downward, and the long hours march along toward inevitable satisfaction of a task accomplished. There is a wellspring of energy from this industry in forward motion that amply provides for all who enter its sphere. When all is said and done, I think I have Lyle Clark to thank the most—for shoving me off the fence.

Very special thanks are also due to Bernie Wilner, Program Manager at Rockwell International's Information Systems Center, who so graciously shared his experience and success with the COMNET installation, and to Ken Asten of Ken Asten & Associates, Inc., and Academic Consultant to the Dean in the MBA Telecommunications Program at Golden Gate University in Los Angeles, who agreed to review the manuscript.

Others who shared invaluable information were Larry Arredondo, author of *Telecommunications Management for Business and Government*, published by The Telecom Library; Carol Martin and Gib Piory at GTE's magnificent Telecommunications Center in Santa Monica; Lissa Zanville of Pacific Telephone; and Henry Wieland, Jr., of USITA. Others who gave me first-hand looks at industry applications were Walt Pepple of Com Systems, Barbara Taylor at Coca Cola, Tracy Lundy of the Mattox Group, and John Pattison of Tel-Analysis Corporation.

Finally, most enduring gratitude goes to my boss, Alan Mentzer, who provided understanding and support while I juggled my writing with my job.

If there is anything at all to regret, it is that no words are adequate to express my thanks to my wonderful son, David, who is so very good to his mother.

THE OFFICE TELEPHONE

A User's Guide

Far Speaking: From Smoke Signals to Fiber Optics

"Are you there, Grandpa?" An eager 4-year-old spoke into the telephone receiver in the kitchen.

From another kitchen 2000 miles away, Grandpa replied, "No, I am *here*. *You* are there."[1]

People have been communicating without being there for many centuries, but it has been just over 100 years since the voice of Alexander Graham Bell was first transmitted electrically. In those 100 years, the telephone has become so commonplace that it almost goes unnoticed, but so necessary that few birthdays can be celebrated without it and vast areas of commercial endeavor depend on it.

■ BEFORE THE BELL

Literally translated from its Greek roots, *tele (far) phone (speaking)* describes an infinitely useful communications tool. Methods for communicating over long distances have been sought since the beginning of

[1]From a Bell System television commercial.

time. Ancient Persians stationed men in towers to shout messages along strategic routes. The House of Taxis operated a messenger service that covered Europe from the fourteenth to the nineteenth century. In 1628, its 20,000 men in blue and silver uniforms delivered messages to and from nobles and merchants. Coded messages have been transmitted by drums, fires, and smoke signals.

The word *telephonte* appeared in a French play as early as 1683, although early French efforts at "far speaking" actually involved signals. The French army was able to send dot-dash-type signals a distance of 6 miles by forcing compressed air through trumpets. By the middle of the nineteenth century, the French military had established a network of towers used for sending visual signals with arm positions and flags. These semaphore towers were considered a vital part of French national defense. The visual semaphore signals are still in use today.

Great Moments in Communication

IT IS 1891 and in France these French soldiers are transmitting voice messages from the ground to the airship above . . . an astonishing distance of four kilometers!

Drawing reproduced courtesy of *Communications News* magazine, a Harcourt Brace Jovanovich publication.

Voice communications are, of course, more desirable. As early as 1644, an English physicist, Dr. Robert Hooke, suggested that sound could travel through wire as well as air. Hooke's device has been called the *string telephone*. In the 1700s, a German student proposed a system of "speaking tubes," which would transmit voice by megaphone along existing roadways. In an article entitled "Electrical Telephony" and published in 1854, a Frenchman named Charles Bourseul suggested that it was not too far-fetched to consider transmitting the spoken word by electricity. Although he never built a device, he described a concept that resembled the telephone as we know it today.

A German professor, John Philip Reis, is credited as the first actually to use electricity for transmitting sound, in 1861. Professor Reis hollowed out a beer-barrel bung (the stopper for the hole in the top of the barrel) and covered it with a sausage skin for his first transmitter. For his first receiver, he coiled a wire around a knitting needle and placed it on a violin as a sounding board. Eventually Reis developed the device to the point that musical sounds could be transmitted well enough to attract audiences, but he was unable to transmit speech successfully.

Other interesting attempts have been recorded. In 1753, Stephen Gray, an Englishman, conveyed electricity by packthread, which is merely a strong, fine twine. A short time later, an article signed only C. M. described a method for sending communications by electricity along a wire.

Nearly a century later, in 1837, Professor C. D. Page, of Salem, Massachusetts, was able to cause sound emission when an iron bar was magnetized and demagnetized. An Italian, Antonio Meucci, applied for a patent on a device for transforming sound waves into electrical current, but the patent was never granted.

Dr. S. D. Cushman, of Racine, Wisconsin, did not attempt to patent his "electrical talking box," which he developed in 1851. Two other inventors did attempt to obtain patents, however. A Dr. Everett, of New Orleans, was issued a patent in 1868 on a device reported to have transmitted audible sounds over a mile of insulated wire. In 1876, J. D. McDonough, of Chicago, attempted to patent his "telelog." Neither of these devices is considered to have influenced the development of the telephone as we know it today.

Significant inventors who were working on telephonic devices during this period were Professor A. E. Dolbear, whose telephones were actually produced and sold by the Western Electric Manufacturing Company (the same company that now manufacturs Bell equipment),

Daniel Drawbaugh, an eccentric Pennsylvanian who carried claims for patent rights all the way to the Supreme Court of the United States, and Elisha Gray, who filed a caveat (notice of invention not yet completed) with the patent office on the same day Alexander Graham Bell filed the patent application for his telephone.

■ THE TELEGRAPH

While Bell and others were still tinkering with inventions for speech transmission, Samuel Finley Breese Morse built the first practical, long-distance communications apparatus, the telegraph, and devised a system of signals, the Morse code. He erected the first telegraph wire in 1838 at Morristown, New Jersey, in a 3-mile circle around the Speedwell Iron Works. This iron works belonged to the family of Theodore N. Vail, who later became the most influential figure in establishing the telephone company as we know it today.

By 1866, the first permanent transatlantic telegraph cable was in operation. By 1876, the year of Bell's patent, some 214,000 miles of telegraph wire were delivering almost 32 million telegrams through 8500 telegraph offices in the United States. The telegraph was clearly the established method for far-speaking, and the invention of the telephone commanded relatively little attention.

■ BELL AND HIS INVENTION

When he invented the telephone, Alexander Graham Bell was working on an improvement for the telegraph, a so-called harmonic telegraph that would send several messages simultaneously over one wire. Bell was, by instinct, an inventor. However, by profession he was a professor of elocution (forerunners of today's speech therapists) at Boston University.

Two of his students, both deaf-mutes, were instrumental in the progress of his telephone invention. The first, 5-year-old Georgie Sanders, was the son of Thomas Sanders, who provided Bell with a place to live and private working space in his cellar. The second, Mabel Hubbard, whom he later married, had lost her hearing due to scarlet fever in infancy. A lovely girl of 15, she provided a great deal of encouragement and captured Bell's heart. She also had a father with financial resources.

Thomas Sanders owned a leather business, and Gardiner Hubbard was a Boston attorney. Both were backers of Bell's invention and his early business ventures, and both were destined for fortunes.

Bell himself combined an understanding of elocution and electricity for the first time in history. Nonetheless, his search for the successful transmission of words over wires was a long and arduous one.

In a way, it started when he was still a boy in Scotland. He and his brother Melville (whose death as a result of tuberculosis later prompted the family's immigration to Canada and eventually to the United States) built a "speaking machine." They used a lamb's larynx given to them by a butcher, along with such items as dental materials, soft rubber, and pieces of cotton to fashion jaws, teeth, tongue, pharynx, and nasal cavities. They even added a wig as a final touch.

With practice, Alexander was able to manipulate the lips, palate, and tongue, while Melville blew into the machine through a tube. Their machine said "Mama" clearly enough to bring a neighbor to see what was the matter with the baby! The so-called speaking machine was a game of boys, but even this early understanding of speech contributed to the invention of the telephone.

It was in the summer of 1874, when Bell was tutoring Georgie Sanders and Mabel Hubbard in Salem, Massachusetts, that he began the first in the series of steps that actually led to the telephone. Using a dead man's ear, he constructed a phonautograph machine, which transcribed the vibrations of the ear drum through a piece of straw into etchings on a piece of smoked glass. The etchings provided valuable information about the transmission of words, but the dead ear prompted Sanders and Hubbard to admonish Bell to get back to work on the harmonic telegraph and abandon his "ear-toy."

Bell then transferred his work to the Boston shop of Charles Williams, who intended to manufacture the device Bell finally invented. Williams also provided Bell with an assistant, Thomas A. Watson.

On a hot day in June of 1875, a metal reed on the harmonic telegraph stuck. Bell, who was in the other room, heard the reed snap when Watson loosened it and realized that the sound had been transmitted electrically over the wire. In that instant, the telephone was born.

However, it was not until nearly a year later—on March 10, 1876—that those famous words were heard over the wire. Bell spilled some acid on his clothes and called, "Mr. Watson, come here. I want you." Watson clearly heard the call in his receiver, and far speaking was a reality.

The first telephone. Photo reproduced with permission of AT&T.

Bell had filed his patent application on February 14; quite coincidentally, patent 174,465 was issued on March 10, the very day Watson heard Bell's call. That patent was later to be called the most valuable patent in the world.

■ A NEW TOY

The bell was ringing, but the phone was not quickly answered. In the beginning, the miraculous potential was not easily comprehended. In fact, the telephone was a subject of considerable ridicule. The speaker was required to shout in a most undignified manner. The quality of sound reproduction was distorted and cluttered. Most significant, Americans simply found it hard to believe that speech could be changed into electricity, sent over wires, and changed back at the other end of the line. Some even thought it was evil. At this point, Bell, Watson,

Sanders, and Hubbard had no way of knowing they held a valuable patent.

Gardiner Hubbard offered to sell the patent rights to Western Union sometime late in 1876 or early in 1877. William Orton, then president of Western Union, turned down the offer, dismissing the telephone as an electrical toy.

Bell took his toy on lecture tours because he needed the money. His demonstrations with musical performances were the most popular. The only practical value people saw in the device was for transmission of music and news (Marconi did not invent the radio until 1909). In his lectures, however, Bell always predicted a central office switching system with suspended or underground cables connecting points throughout the city and even throughout the country, but few in his audiences were able to envision themselves talking on telephones.

The first telephone exchange opened in Boston on May 17, 1877. It connected the Charles Williams Company (where the invention took place), the Shoe and Leather Bank (Sanders was in the leather business), Brewster, Basset and Company (bankers), and E. T. Holmes (who held the license to operate the exchange). Holmes was in the burglar alarm business, and he saw the telephone as a useful adjunct at night to his alarm system. Later, Holmes installed the telephone in several express companies, where it proved its first real commercial value in receiving and transmitting orders.

Still, it was slow to shed its image as a toy. The telegraph was the accepted medium of communication, and the task of establishing the telephone as its equal was overwhelming. Gardiner Hubbard was a staunch campaigner, however, spreading the word to all who would listen. In 1877, he and his three colleagues prepared a document claiming the telephone's superiority over the telegraph for the following reasons:

1. No skilled operator is required, but direct communication may be had by speech without the intervention of a third person.
2. The communication is much more rapid, the average number of words transmitted in a minute by the Morse sounder being from fifteen to twenty, by telephone from one to two hundred.
3. No expense is required, either for its operation or repair. It needs no battery and has no complicated machinery. It is unsurpassed for economy and simplicity.

This document could have been a model for selling many business machines on the market today, but it had little effect on businesspeople at that time.

It took a disaster rather than a document to give the telephone its first public appeal. A railway accident occurred in January 1878 near Tariffville, Connecticut. The Western Union operator notified the Western Union office at Isaac Smith's drugstore, which fortunately was connected by telephone to 21 physicians. Smith's calls enabled those doctors to get to the scene with medical supplies in time to save lives that would otherwise have been lost. The story was widely publicized.

The consumer of the day was even less responsive than the businessman of the day. Ponton's Telephone Central Service of Titusville, Pennsylvania, issued an advertisement saying:

> The system is extremely simple. All parties who wish to adopt it must have separate wires from their house, office, factory, hotel, store, bank or restaurant to a central switch room where any one wire can instantaneously be connected with any other wire. Supposing that one hundred persons adopt this system, and that the average length of each wire is half a mile, it would give each person the privilege of using fifty miles of wire at less cost than it could be done with only one mile in the private line.

After this tedious description of a telephone network, the circular went on to describe the practical value of the telephone:

> In domestic life the telephone can put the user in instant communication with the grocer, butcher, baker . . . [a list of 176 businesses and services followed] . . . and other places and persons too numerous to mention.

The Ponton Company went out of business.

In 1879, the Pittsburgh telephone directory listed 300 subscribers, only 6 of whom were residents. Even these were proprietors of businesses, who wanted to be in constant touch with their offices. Most of the early subscribers were banks, hotels, and doctors, although the idea of private conversations was beginning to appeal to some of the captains of industry who protected secrets in that era of vicious competition. In their wildest imaginations, none of them could have foreseen the scope of the telephone's role in business today.

■ BRINGING UP A GIANT

When Bell and his three backers began to see enough business to justify forming a company, they drew an agreement giving Bell, Hubbard and Sanders 30 percent each and Watson 10 percent. The organization was known as the Bell Telephone Association, but the value of its stock was considered a joke.

Nonetheless, it was not long after President Orton brushed the telephone aside that one of Western Union's subsidiaries reported a telegraph machine had been "superseded" by a telephone. That slight hint of competition awakened the sleeping giant and marked the beginning of a tremendous struggle between the little telephone association and the behemoth, Western Union.

The giant telegraph company waged its campaign on two fronts: It established the American Speaking-Telephone Company in an attempt to drive Bell out of the infant telephone industry, and it presented a formidable challenge against Bell's patent.

The announcement of Western Union's entry into the telephone business caused just the opposite of its intended effect. Instead of stamping out the little Bell Telephone Association, the announcement sent them a flood of business. Now Gardiner Hubbard was faced with overwhelming demand instead of discouraging rejection. The situation was critical. Realizing that the association had absolutely no capital (and about the same income) and that none of the four partners had the knowledge or experience to manage a rapidly growing firm, Hubbard offered the position of General Manager to Theodore N. Vail at a salary of $3,500 per year. Vail, eager for the challenge, reported for work a week later.

Few men in history have been as well qualified for such a task. Vail had been involved with telegraphy since his childhood, when Morse erected the first telegraph wire at his family's ironworks. His principal occupation had been with the Government Mail Service. When Bell's first patent was issued in 1876, Vail was head of the entire nation's mail service and had completely reorganized it. Most important, he had been in frequent—though largely coincidental—contact with Hubbard and was totally sold on the potential of the telephone.

Theodore Vail was a man of vision. From his experience with the telegraph and the postal service, he had acquired a firm grasp on the kind of organization required for a communications service to cover an entire nation. He was hard-driving and courageous, and he was not afraid of the giant, Western Union. For his first task, he guided the company successfully through its early, lean years of bitter competition against Western Union. Then, with capital of $850,000, he reorganized the company as the National Bell Telephone Company.

As soon as success was apparent, claims against the patent came from everywhere. The most serious was the one backed by Western

Union. With skill, finesse, and a competent staff, Vail steered the course to a landmark agreement between Bell and Western Union in November 1879.

This agreement, in effect, required Western Union to get out of the telephone business and recognize Bell's patents. In return, Bell was required to purchase Western Union's telephone system and stay out of the telegraph business. As a result, the way was paved for the National Bell Telephone Company to take its place as a giant beside Western Union.

With $6 million in capital, Vail reorganized once again the following year as the American Bell Telephone Company, and Alexander Graham Bell's telephone had reached a pinnacle of success.

It was at this juncture that the inventor and his original backers went their separate ways. Hubbard joined the National Geographic

Mr. and Mrs. Alexander Graham Bell. Photographed by Gilbert Grosvenor and reproduced with permission of AT&T.

Society, Sanders bought a gold mine in Colorado, and Watson went into the ship-building business. Bell had given all his stock to Mabel Hubbard on their wedding day, and he returned to other inventions and his work teaching the deaf.

■ VAIL AND HIS GIANT

Vail was now alone on a threshold. There were no telephone companies whose lead he could follow. There were no books or courses for him to study. He and the men he hired had to respond to rapidly increasing customer demands with courage and innovation.

Other patent struggles continued briefly, but the biggest demands were for technological improvements and expanded service. Pioneers at the American Bell Telephone Company had to meet these demands at a time when electrical engineering had not yet even become a profession.

This maze of wires in Pratt, Kansas, in 1909 was only a hint of the coming impact of telephones in the Southwest. Photograph reproduced with permission of AT&T.

The quality of the sound had to be improved. This was accomplished by using iron for the diaphragm instead of goldbeater's skin. Wires had to be strung everywhere to meet growing demands for service. An organized plan was needed. Eventually, overhead wires filled the skies, and no more could be added. Wire substantial enough to be buried had to be developed.

Switchboard systems had to be perfected. As subscribers grew in number, switching systems became inadequate, and a totally new approach had to be designed. The activity in the switching rooms was chaotic, and connection time was unreasonably long. Techniques had to be developed and personnel had to be trained to handle calls quickly and politely under heavy pressure.

When wire was strung west of the Mississippi, technical problems required special ingenuity. Indians wanted to use the bright copper wire for earrings. Bears thought the humming sound was made by bees and constantly gnawed down the poles in quest of imagined honey.

Through it all, however, Vail clearly saw the real far-speaking tool. He pursued his vision of a national communications system that could provide faster and more direct service than either the telegraph or the mail. In a letter he wrote in 1879, he said, "Tell our agents that we have a proposition on foot to connect the different cities for the purpose of personal communication, and in other ways to organize a 'grand telephonic system.'" To this end, he formed a central company in 1885, linking all the local companies together. The company was called American Telephone and Telegraph Company, and its stated purpose was "to connect one or more points in each and every city, town, or place in the State of New York, with one or more points in each and every other city, town, or place in said State, and in each and every other of the United States, and in Canada, and Mexico; and each and every of said cities, towns, and places is to be connected with each and every other city, town, or place in said States and countries, and also by cable and other appropriate means with the rest of the known world."

Vail set out to prove his purpose was not just a dream. By 1892, half the people in the United States were within talking distance of one another. Largely as a promotion effort, lavishly decorated long-distance "salons" were opened. Customers were transported to and from these salons by cab. Making a long distance call was like going to the theater, and Vail's "grand telephonic system" was taking shape. American Telephone and Telegraph Company—AT&T—was clearly on the horizon.

In addition to laying the foundation for his grand telephonic system, Vail planted the seeds from which a totally new industrial philosophy grew. He created an innovative company and a unique industry, both of which still exist today. First he saw the need for standardization. By 1881, Vail had bought all the manufacturers of telephone equipment and consolidated them into one company, the Western Electric Company. Next, by showing competition in the industry to be wasteful duplication, he established the premise that there should be only one telephone company. Competing telephone companies required customers to purchase several telephones to reach different areas. Vail made the idea of a natural monopoly a reality and opened the door to government regulation.

In the early 1900s, independent telephone companies sought to be included as members of the Bell family, and Vail made the first reference to the Bell System, existing for a higher purpose than profit. Vail had conceived the first privately owned corporation organized specifically to serve the public interest.

By the 1960s and 1970s, AT&T was employing nearly a million people and Americans were making more than 250 million telephone

Management of AT&T's nationwide network through control centers such as this one assures efficient use of the network around the clock. Photo reproduced with permission of AT&T.

calls every day. A telecommunications satellite was orbiting the earth. In the early 1980s, an electronic switching system installed in the Midwest was capable of processing 550,000 calls every hour. The grand telephonic system had covered the planet, reached into space, and entered technological arenas beyond the imaginations of both Alexander Graham Bell and Theodore Vail.

■ AMERICAN TELEPHONE & TELEGRAPH COMPANY

Until the 1982 Consent Decree, AT&T was the largest corporation in the world. It has been affectionately called Ma Bell and defiantly called a monolith. It has been studied and criticized as a model for corporate management and as a wonder of the Industrial Age. It has, with justification, been both applauded and condemned by the public it serves. Even Ma Bell's strongest critics, however, concede that the "Bell system" provides the best telephone service in the world.

Before 1982, AT&T was the parent of the 22 telephone companies serving specific geographic areas of the United States. However, the 1982 Consent Decree required AT&T to give up ownership of these operating companies. Although they will remain interconnected with the long-distance network that is controlled by AT&T's Long Lines Department, they are now separate companies, and they are expected to announce significant changes when they reorganize into regional divisions in 1984, which are shown in the back of the book.

The newest additions to the Bell System are AT&T Information Systems an unregulated subsidiary of AT&T offering shared data communications network services, and Advanced Mobile Phone Service, Inc., which offers mobile phone service using a new technology called *cellular radio.*

AT&T is still the sole owner of the Western Electric Company, which manufactures all the equipment that Bell System companies provide their customers. If it were not wholly owned by AT&T, Western Electric would, by itself, be the twelfth largest industrial company in the United States.

AT&T and Western Electric together have owned the Bell Telephone Laboratories since 1925. Between 1925 and 1975, no fewer than 18,000 patents were issued to Bell Lab scientists, one for every working day in that 50-year period. Several of these inventions are particularly significant in the role of the telephone in today's office. In the 1930s, Bell Lab scientists developed and produced the coaxial cable, which permitted simultaneous transmission of many signals. In the early 1950s, Bell Labs introduced microwave radio-relay systems, which permitted wireless transmission of signals at much greater capacities. In the late 1950s and early 1960s, increasing traffic demands necessitated the development of computerized telephone exchanges. Without these three developments, the spectacular volume of calls made in the 1970s and expected in the 1980s and 1990s could not be handled.[2]

Other telephone-related inventions have contributed to areas outside the telecommunications industry as well. The transistor, which was invented in 1948 and for which three Bell Lab scientists were awarded a Nobel Prize, contributed to the development of the computer industry, to radio and television, and to space missions. The invention of the transistor has been given importance equal to that of the telephone itself.

More recently, Bell Labs invested about $50 million in research connected with Telstar I, the first communications satellite, which was launched in 1962. AT&T financed the $1 million cost of the satellite and the $3 million cost of the launch. This satellite and others that followed are now operated by Comsat, a separate corporation created by the Communications Satellite Act. Bell Lab scientists have also contributed to the development of the laser and fiber optics. Laser technology is being used in business applications for microfilm records storage and high-speed document printers. Fiber-optics technology is contributing to improvements in office copiers. Both lasers and fiber optics promise vastly increased transmission capacities and speeds for the telecommunications industry in the immediate future. The energy-saving potential of fiber optics is even more significant; about a thousand times as much energy is required to produce copper wire as is required to produce optical fiber.

▪ THE INDEPENDENT TELEPHONE COMPANIES

The Bell System companies are by no means the only telephone companies in the United States. There have always been independent telephone companies serving specific geographic areas. They do not compete against Bell companies; rather they interconnect with AT&T's

[2]For an excellent study of American Telephone and Telegraph Company see John Brooks, *Telephone: The First Hundred Years* (New York: Harper & Row, Publishers, 1975-76).

Long Lines Department to provide long-distance services for their subscribers.

Today there are over 1400 independent companies operating in all 50 states. Combined, they control 20 percent of the nation's telephones in a little more than half of the geographic area of the United States. Many are small companies serving rural areas or small towns. One has as few as 40 telephones. Another—in Bryant Pond, Maine—was still using telephone instruments housed in oak boxes and actuated with cranks as recently as 1982. All calls for this independent were switched by operators at a switchboard located in the living room of the family who owned the company. Residents of that small town felt their telephone system met their needs adequately and raised a protest when it was sold to new owners, who wanted to modernize the equipment.

Most of the independents belong to the United States Independent Telephone Association (USITA); some of the larger members of this

GTE's COMSTAR satellite under test. Photo courtesy of General Telephone Company of California.

group are listed in the back of this book. The largest of the independents is General Telephone and Electronics (GTE), which serves over 14,000,000 telephones. GTE has both operating companies and a manufacturing subsidiary, GTE Network Systems, formerly Automatic Electronic Company and the GTE Lenkurt Electrical Company. GTE also owns GTE Telenet, a company that offers data-transmission services and, GTE Satellite Systems.

■ TELECOMMUNICATIONS AND COMPETITION

As strong as the independents may be, however, any discussion of the telephone industry is dominated by AT&T and characterized by the degree of competition in the marketplace. From about 1910 until 1968, AT&T was permitted to enjoy a so-called natural monopoly, with varying degrees of government regulation. (Independent telephone companies are also monopolies in the geographic areas they serve.) The Interstate Commerce Commission (ICC) was the regulating agency from 1910 until 1934, and the Federal Communications Commission (FCC) has been the regulating agency since 1934.

In natural monopoly, a company is in an industry in which the best job can be done at the least cost by one firm having no competitors. Similar monopolies exist in other utility industries, such as water, power, and public transportation. In the telecommunications industry, however, recent FCC decisions have permitted competitors to enter the marketplace. This new competition has altered the nature of business decisions concerning telephone equipment and long distance services.

Before 1968, nearly all business telephone equipment in use was manufactured by Western Electric, sold to the Bell System companies, and *leased* to users. Beginning with the famous Carterfone Decision of 1968, it became legal for other firms to manufacture, sell, and install telephone equipment and interconnect it to Bell System lines. Most of these new competitors make equipment for business users and will *either* lease or sell to their customers. This means the business firm can now own telephone equipment chosen from among vendors competing in the marketplace. These equipment manufacturers and vendors have created a new industry called the *interconnect industry*.

The second change in regulation involves long-distance services. Several decisions in the 1970s permitted similar interconnection by firms known as Specialized Common Carriers (SCCs), who have built networks using microwave transmission, for the most part, and connecting primarily larger cities. The competitors entering this marketplace offer both business and residential users alternative services at lower rates for high volumes of calls.

Both the interconnect companies, who offer equipment, and the SCCs, who offer carrier services, bring competition into the telecommunications industry at a time of rapid change and advancing technology. Far speaking has become an extremely complex activity, especially in the office. For business managers, both the role and cost of telephones are increasingly significant to the success of every phase of the operations. Alternative equipment and services mean that every manager, and indeed every employee, must know more about the telephone than ever before.

Northern Telecom's SL-1 computerized switch and telephone equipment. Photo courtesy of Northern Telecom, Inc.

QUESTIONS FOR REINFORCEMENT AND DISCUSSION

1. In what way are the following words related to the telephone (communication over distance by electronic signal)? Use a dictionary if necessary. (*Hint*: Some aren't!)

teledu	telepathic
teleconference	telephoto
telecopier	teleprocessing
telegraph	teletex
teleman	telethon
teleost	teletypewriter

2. Do you remember when you learned how to use the telephone?
   a. How old were you?
   b. Who taught you what to do?
   c. What did your telephone look like?
   d. What was your telephone number?
   e. What were you taught to do?
   f. What kind of telephone calls did you make?
   g. What would have been different about your life if you did not have a telephone then?
   h. Did you learn enough then to use a telephone in an office now?
3. How would life be different today if there were no telephones in our homes? In our offices?
4. How do you think you would have reacted to the invention of the telephone if you had been alive in 1876?
5. How would you go about starting a business if you invented the telephone today?
6. Can you think of any recent inventions or discoveries that may turn out to be as important as the telephone?
7. Do you think the telephone company should be a monopoly?
8. Do you think the government should regulate the telephone industry?
9. How do you feel about the telephone company?
10. Are you aware of any changes in the telephone or telephone services that have occurred recently?

SPECIAL PROJECTS

1. Go to the library and study some aspect of the history of the telephone, AT&T, or the telephone industry. Write a brief paper on your findings.
2. Check both the Yellow Pages and the White Pages of your telephone directory and make a list of the firms and association you find listed under *telephone, telegraph, telecommunications, communications equipment, or communications consultants*. Contact several of them to find out the nature of their

businesses, their activities, their employees, their customers, and their products or services. Report your findings to the class.

3. Go to the library and find out more information about an invention or technological development related to telecommunications or for which Bell Lab scientists were responsible. Write a brief paper describing it and its role in telecommunications, or report your findings to the class.
4. Locate a large firm near you that has a central communications department. Contact the manager of the department and arrange for a visit or a telephone interview. Find out the type of equipment they have, the number of telephone users they serve, the type of communications they support, and some of the problems they experience. Report your findings to the class.

2

The Many Roles of the Office Telephone

■ THE TELEPHONE LINKS PEOPLE

Telephone and telecommunication systems find their way into nearly all the activities in the offices of modern business and government. The most familiar roles of the telephone concern its use in conversations and exchanges of information between people. The wheels of business may be powered by profit and the wheels of government may be powered by votes, but those wheels are most certainly oiled when people keep talking.

□ The Telephone Speaks

The telephone instrument has a voice of its own that is separate and distinct from the human who is originating the call. Its voice is usually a repeated ring. Sometimes it is a loud, rasping buzz. Modern instruments may sound beeps, musical chimes, or quiet, warbling tones. Some make no sound at all, but rather speak with a steady or flashing light. Some even display a message in writing. Whatever voice the telephone instrument may use, it is simply saying, "Someone has a message to convey."

For the person who hears that voice, however, it may say any one of a dozen different things, depending upon that person's expectations. For the lovestruck teenager, it may be heaven calling. For the anxious father, it may be a boy or a girl. For the emergency desk, it may be a call to save a life. For the sales manager, it may be the biggest order of the year. For the customer service manager, it may require delicate handling. For the systems manager with a down computer, it may restart the entire office operation. For the proprietor on the brink of bankruptcy, it may be the creditor who can close down the business. For the busy secretary, it may be an intolerable interruption. For the new employee, it may be a frightful uncertainty. For anyone, it may be a complete surprise.

Regardless of the situation, however, the telephone instrument *always* says that someone has a message to convey. It demands immediate and complete attention.

□ The Ubiquitous Telephone

"As near as your telephone" is a familiar distance. It is usually somewhere between an arm's length and a few steps. It is seldom more than a few minutes' travel. Some portable telephones can be carried about freely. The telephone is everywhere, especially for today's business executive. At home the executive may have a phone in the hall, kitchen, den, living room, garage, or by the pool. Most likely, there are several.

This ROLM® ETS™ 100 electronic-key telephone set displays a message telling the user to call extension 2208. Photo courtesy of ROLM Corporation.

There may even be a telephone in the car. On the way to the office, that car passes several public telephones; perhaps there is one near the newsstand where the executive picks up a morning paper.

Arriving at the office, the executive may be cleared by a guard whose telephone is part of the company's security system. Inside, the receptionist is routing a rapid succession of incoming calls on a busy, computerized switchboard console for the company's 200 extensions. The department's administrative-support secretary is screening calls and taking messages for several other executives.

Arriving at his or her desk, the executive may find the phone ringing and ready to determine the first task of the day. The executive may also need to review data received during the night from field personnel and recorded on a cassette tape. Before visiting another office, the executive may program his or her telephone to forward calls automatically to that office. At lunch, the executive may be paged over a phone system so that calls can be received immediately.

Many switchboard attendants report that the first thing businesspeople do when returning to the office is pick up their telephones. So many calls are made during peak periods that some companies employ automatic queuing equipment to answer calls and hold them in line until someone is available to handle them.

So goes the day, and the evening as well. In fact, the company's telecommunications system may continue to accumulate calls and transmit data throughout the night.

Telephones are the only equipment that are found in nearly every business. A store may not have typewriters, but it has telephones. An office may not have cash registers, but it has telephones. A factory may have neither typewriters nor cash registers, but it has telephones. It has been estimated that telephones are used at some point in at least 90 percent of all business transactions. Telephones are management's most familiar tool. In an office, telephones are an important part of everyone's workday, and they must be used with competence and confidence.

□ The Telephone as an Extension

The telephone comes as near as any human invention to being an extension of the human body. The person originating the call may extend only his or her voice across the distance, but that voice is sufficient to accomplish many tasks. It transmits instructions, information, feelings, and influence. On the other hand, just by having the telephone service, that same person extends his or her availability to anyone else who uses the telephone almost anywhere in the world. The businessperson waiting for an important call may feel tied to the telephone.

The telephone invades privacy while at the same time protecting it. The ringing telephone must be answered. It cannot be escaped. The identity and importance of the caller cannot be determined any other way. Once the call is answered, the caller takes precedence over all. To hang up before the message is completely conveyed is extremely rude. Another person present in the room or office must wait for the person on the phone, no matter who that caller may be. The precedence and instant accessibility created by the telephone can be both helpful and harmful in an office. The telephone as an extension of the human body possesses a unique importance and necessitates special techniques for achieving good human relations.

□ Speaking Freely

The telephone as an extension of the human body also enables people to speak more freely than they would with other means of communication over distance. Paper inhibits expression for most people because writing is slow, document production often requires typing, and the

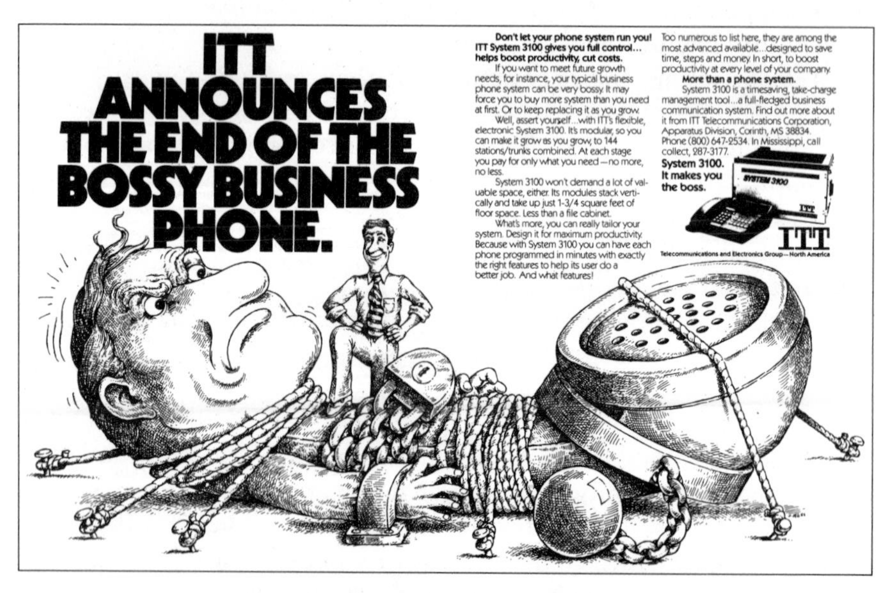

Reproduced from a magazine advertisement for the System 3100 computerized phone system with permission of ITT Telecommunications.

printed word can create a binding commitment. The mere thought of speaking before groups or via broadcast media causes many people to all but forget how to talk.

For nearly everyone, however, talking on the telephone is as easy and natural as talking across a table or across a desk. People can keep talking, enjoying the luxury of knowing that their ideas can be completely understood exactly as they are meant with no possibility of miscommunication.

☐ First Impressions

Many people get their first impression of an office or even of an entire business on the telephone. The caller does not know whether the office is under construction or under control. Whatever the existing situation may be, the caller assesses the company, its management and employees, and its products and services by the way a call is answered and handled. First impressions are important. The techniques for assuring good human relations and good first impressions with every telephone call will be discussed in a later chapter.

☐ Internal Communications

Nearly two-thirds of telephone usage in many offices is internal. The role of the telephone in internal communications is to facilitate the flow of work. Some companies have special telephone lines for internal communications, known as *intercom* or just *com* lines. These lines may be used for routing incoming calls, conveying messages between people, and exchanging information between offices and departments.

An intercom line, however, is just one of several alternatives for interoffice communications, and the method of interoffice communication must be chosen wisely in order to achieve efficient workflow. Since the intercom must be used by a number of people, it should generally be kept open as much as possible.

Here are some situations in which the intercom line may not be the best choice:

1. A company with a high volume of incoming calls may find the intercom line inadequate for serving both the receptionist, who is routing the calls, and the rest of the office staff. Such a firm may choose automatic call-routing equipment for incoming calls instead. Equipment and systems for this purpose are discussed in a later chapter.

2 A routine task involving an exchange of available information readily may be handled by a note or interoffice memorandum sent through the company's internal mail system. This paper flow exchange, sometimes called the *in-basket approach,* allows everyone to prioritize their work loads, rather than requiring one worker to interrupt critical activity to handle a routine request for another worker.

3 For a project involving a lengthy exchange of information or discussion, a face-to-face meeting may be more efficient than a telephone call. A long call may tie up a com line when it is needed by others. If the project requires several people, a prearranged meeting may be necessary to insure the availability of everyone involved. Availability can be a very scarce commodity in a busy office.

When a relatively small amount of information is to be exchanged or an immediate need must be filled, the internal telephone line is the best. The telephone can demand immediate attention, but it cannot by itself create efficiency. Only people using good common sense can determine when the telephone is the right tool for the right job.

☐ External Communications

While the role of the telephone in internal communication is largely to promote efficient flow of work, in external communication it can be the difference between the success or failure of the business. More business information flows through telephone lines today than ever before. Information has taken on a new importance with the advent of computers.

The telset in the foreground is the 21-line answering position of the Com-Key (TM) 2152. Photo reproduced with permission of AT&T.

Managers give increasing importance to a function called *information processing*, which is discussed in the next section of this chapter. Sociologists tell us we are entering a third stage of civilization, called the Information Age. An understanding of the three stages of civilization can help illustrate the role of the telephone in business and society today.

The first stage is called the Agricultural Age; it is described as the era when people stopped hunting and foraging for food and began cultivating land. The plow and the wheel are said to have given birth to the Agricultural Revolution. In this era, power rested in the hands of those who owned the land.

The second stage is recorded in history books as the Industrial Age, when energy and technology promoted production and selling of goods and services in a much more complex society. The steam engine is said to have given birth to the Industrial Revolution. In this second era, power rests in the hands of those with resources of energy, methods of production, and technology.

The third stage is being called the Information Age, in which power will rest in the hands of those having accurate and timely information.[1] To be successful, a company must be able to manage information efficiently and effectively. This information is the product of the office. The computer gave birth to the Information Revolution, and the telephone—through which information flows almost instantaneously—is playing a significant role in its development.

In business firms and government agencies, the way by which the telephone facilitates information flow may vary from one department or activity to another. For example, top managers require instant access to data concerning current market information, inventory levels, production schedules, and cash flow. Officials must also verbally communicate directives to subordinates, who may be located several doors away, several floors away, on another side of the city, or on another side of the world.

The telephone is the most practical and effective method of communication for some specific business and service functions.

- Customers can inquire about products and services more conveniently by telephone than by a visit.
- Routine orders can be placed more quickly by telephone than by mail.

[1] For a thorough study of the Information Age, read Alvin Toffler's *The Third Wave* (New York: Bantam Books, Inc., 1980).

- Some sales promotion efforts are more effective by telephone contact with specific individuals than by advertisement to the general public in newspapers, magazines, or on radio and television.
- Collection of overdue accounts can be handled more firmly and discretely by telephone than by mail.
- Public information services can reach the homebound and people in remote locations who could not be served without telephones.
- Huge amounts of data can be transmitted across continents in a few seconds by telecommunications facilities.
- Transactions can be made just as swiftly in international markets as in local markets because of the telephone.
- Finally, the telephone often plays a critical role in national defense and in diplomatic relations between friendly, as well as unfriendly, nations.

These are just a few of the roles played by the telephone in the external communications of a business or government office. As shown in this and subsequent chapters, these and other roles of telecommunications are constantly changing and broadening in scope.

□ Hot Lines

The telephone helps to disseminate a great deal of useful information through both business and public service agencies. Some of this information merely serves useful purposes in daily living activities, and some plays life-giving and lifesaving roles in the human drama. The popularity of *hot lines* in the past several decades illustrates some of these roles.

Recorded messages enable people to dial the current weather information, accurate time, road and traffic conditions, and the schedule at the zoo. Services such as Dial-a-Ride, Dial-a-Joke, and Dial-a-Prayer have been offered.

In some places, trained counselors are available 24 hours a day for suicide prevention, crisis intervention, and family counseling. Hot lines are open for special groups such as teens, the elderly, drug abusers, alcoholics, gamblers, and overeaters. Some service centers maintain open lines for special problems such as rape, unwanted pregnancy, poison information, rumor control, stock market trends, and even questions about grammar and punctuation. Law enforcement agencies can open safe and anonymous channels of communication for people who want to report information about crimes, child abuse, consumer fraud, and income tax evasion.

■ THE TELEPHONE LINKS MACHINES

Alexander Graham Bell and his fellow pioneers probably had a fairly clear vision of the communications between people that would be possible with the telephone, but they probably had little inkling of the role telecommunications would play as the link in the integrated systems appearing in business and government offices today and predicted for the future. The term *telecommunications* refers to telephone lines as movers of information. Voice is not the only form in which information can be transmitted electronically. Telephone paths are also used to transmit data (numbers) and images (facsimile and video). Such communications occur from one machine to another. In other words, machines—as well as people—are talking on the telephone.

Communicating machines permit transmission of large volumes of information at high speeds over short or long distances. For example, the entire text of *War and Peace* can be transmitted from a city in the United States to a city in South Africa in only 1 second. Communicating machines also permit information entered or stored in one machine to be transmitted to another machine for processing or reformatting. This means, for example, that numbers stored as electronic signals on magnetic tape in a computer can be transmitted through telephone lines to a document printer on another floor and typed in columns on paper (known as *hard copy*) entirely without human intervention.

□ Information Processing

A discussion of the functions of an office and how these functions are being automated helps to illustrate the role of telecommunications in information processing. The automated functions of information processing may be categorized as follows: data processing, word processing, reprographics (photocopying and printing), storage and retrieval (filing), and distribution. These functions are not new to the office, but machines that perform them are making rapid and significant changes in office activities.

Data Processing. Computers and data processing equipment have been used in business and government for several decades. As microcomputers and minicomputers become smaller and cheaper, they are found more and more frequently in both large and small offices. Some offices with large, centralized computers are adding smaller computers in remote locations and distributing data processing functions to individual departments. Telecommunications networks enable these computers to communicate instantly over short and long distances (across a hall or across an ocean) by telephone lines, microwave relay stations, and satellites. Such systems are known as *data communication systems*.

Word Processing. A *word processing machine* is a typewriter with a memory, sometimes called an *intelligent typewriter*. Word processing equipment enables the secretary to type at rough-draft speeds into the word processor's electronic memory. The stored text can be corrected, revised, and rearranged in the memory until both the machine operator and the author are satisfied with it. The text is then printed out at high speeds with no errors or corrections. Every copy is individually typed, and every copy is perfect. The text can be stored permanently outside the machine on magnetic tape, disks, or cards and printed over and over again with or without additional revisions or updating.

Some word processors have communicating features, so that stored text can be transmitted by telephone lines to word processors in other offices or to other information processing equipment, all without rekeying (typing).

Another important use of the telephone in word processing procedures is in the dictation phase of the document production cycle. Authors can have access to centralized dictation equipment by telephone, which means they can dictate from their offices, their homes, or even from distant cities when they are on business trips.

Reprographics. Reprographics is a term used to describe document copying, whether it be one copy on a small office copier or thousands of copies on a large printing press. Photocopy machines have been communicating for some time. Using telephone lines, machines known as *telecopiers* can photograph documents in one location and transmit the image to a receiver in another location that will produce a photocopy. Today's sophisticated copiers also transmit images to machines other than photocopiers. In addition, some photocopiers can perform certain processing functions on the images, such as reformatting.

When high-quality printing is more desirable for the finished product than typing or photocopying, text that is entered or stored in word processing equipment can be transmitted by telephone lines to photocomposition equipment. Modern photocomposition equipment can accept output from word processors, change the style and size of the type, rearrange text material to allow for artwork, and prepare the text for many types of printing presses. This eliminates a full step in the printing process by making it unnecessary for a typesetter to retype the copy.

Storage and Retrieval. Today's office technologies offer two types of document storage as alternatives to filing paper: (1) magnetic storage in computers or on magnetic media; and (2) micrographic storage on film. Magnetic storage means that images are stored as electronic signals on tape, cards, or disks. This method can be more expensive, but it allows for the greatest amount of storage in the smallest space and also the shortest retrieval time. Micrographic storage means that documents are photographed and stored on film. The film may be in the form of rolls, called *microfilm*; on strips called *microfiche*; or on cards, called *aperture cards*. Microfilm storage involves slower retrieval and requires more storage space than electronic storage, but still far less time and space than paper.

Later, the information stored either electronically or on film can be retrieved in readable form by either one of two methods: it can be displayed on a video screen, or it can be printed out on hard copy. Telephone lines sometimes provide the link between the storage equipment and the screen or printer.

Distribution. The term *distribution* is used to describe any method by which information travels from one point to another. Traditionally, information has been typed on paper and mailed to its destination either through an internal system for mail distribution or the United States Postal Service. However, exploding information requirements are producing a paper avalanche, and paper simply cannot be physically moved fast enough to satisfy demands in the Information Age. Telecommunications offers an alternative method of distribution known as Electronic Document Distribution (EDD), which saves both time and paper.

Electronic document distribution (also called *electronic mail*) requires a system of computers, word processors, and terminals, which are linked by telephone paths. Documents are originated as electronic signals, transmitted from point to point on the network, viewed on video screens, and stored electronically. (Hard copy is printed out only if it is absolutely necessary.) Even companies with worldwide networks can create, distribute, and store messages and documents almost instantly and entirely without the use of paper.

Mailgram is an electronic mail service offered by Western Union in conjunction with the United States Postal Service. Messages are called into Western Union offices by customers, converted to electronic signals, transmitted by Western Union lines to other locations, printed on paper, and delivered to recipients by postal carriers. A similar international service, known as Electronic Computer-Originated Messages (ECOM), has been inaugurated by the United States Postal Service.

☐ Telecommuting

The capability for information-processing machines to use telephones to communicate over distances means that office tasks can be performed from remote locations. In fact, workers with equipment installed in their homes do not have to travel to offices. For example, a computer terminal can be located in the home of a programmer, or a word processing work station can be installed in the home of a correspondence secretary. This idea is called *telecommuting*, and these people are called *telecommuters*.

The costs of both public and private transportation, as well as of space for parking and offices in inner cities, are going up, while the costs of telecommunication and equipment installations are coming down. Therefore, telecommuting is rapidly becoming an attractive alternative for both workers and office managers. In addition, people who cannot conveniently leave home, such as the handicapped or single parents of young children, can join the office work force much more easily.

■ THE TELEPHONE CREATES JOBS

This chapter has discussed some of the jobs done by the telephone and telecommunications. Certainly Alexander Graham Bell's gadget has shed its image as a toy and earned universal respect as an important tool for communication in the Information Age. In addition, the telephone and telecommunications have also created jobs, many of which are in business and office occupations.

Some of these jobs require the equivalent of a high school education or perhaps some vocational training, while others require college degrees and even graduate education. Some require machine operator skills, while others require a great deal of technical and scientific knowledge. All require good human relations skills—the ability to communicate and get along with people well—for that is what the telephone is all about.

Bell System companies and the 1400 or more independent companies, combined with the interconnect industry and the SCC's, make up a significant segment of the American labor force. These firms employ people in all occupations, including ground staff, security guards, nurses, and research scientists, in addition to secretaries, engineers, and corporate officers.

They also employ people in nonoffice jobs that are specifically related to telephone equipment and services, such as people who empty coin boxes or deliver directories, as well as highly trained lineworkers, installers, framepeople, repairpeople, and cable splicers. Some of the specialized technicians and engineers in the telephone industry include central office equipment technicians, traffic study technicians, switching engineers, transmission engineers, and outside plant engineers.

There are several specific categories of business and office occupations related to telecommunications.

☐ Telephone Answering Personnel

Next to that of the typist, the occupation most instrumental in bringing women into the work force was that of telephone operator. The operator's duties include answering calls, routing calls, taking messages, and operating telephone equipment. Telephone operators in firms with complex telephone systems must have a great deal of knowledge about equipment, services, and rates in order to use systems efficiently. In addition, they must be highly skilled at dealing with people. Some examples of operator occupations filled by both men and women include telephone receptionist, switchboard operator, telephone console attendant, information operator, long distance operator, and answering service operator.

☐ Marketing Personnel

Some salespeople do all their work over the telephone. They may be called telephone solicitors, telephone order clerks, or telephone adtakers.

Companies in the telecommunications industry employ specialized salespeople, who assist customers and users when they acquire tele-

This telephone receptionist handles hundreds of calls every day using the Wescom 580 PBX attendant console with busy lamp field. Photo courtesy of Rockwell International Corporation, Switching Systems Division.

These telephone operators are routing calls on an electronic tandem switching system (ETS). Photo reproduced with permission of AT&T.

phone equipment or subscribe to services. These marketing personnel must often possess extensive technical knowledge about equipment, services, systems, and rates. They may be called telephone customer service representatives, market representatives, or market support representatives.

□ Communications Personnel

The operation of telecommunications networks and systems and even some telephone equipment has long been sufficiently broad in scope to justify an entire occupational category requiring special knowledge and training. Titles for these occupations may include communications manager, communications supervisor, communications analyst, rate analyst, and communications specialist. More recently, data communications operations require personnel with knowledge of both data processing and telecommunications. Persons in these occupations are called data communications analysts, data communications specialists, and data communications managers.

□ Consulting Personnel

Telephone companies employ people whose job is to advise customers about equipment, services, systems, and costs. These people are called residential service representatives, commercial service representatives, and customer support representatives.

Business and government offices with very large and specialized requirements for communications may hire outside consultants or consulting firms to help them design their systems, choose both vendors and equipment, work out problems during installation and start-up, and evaluate and maintain their systems. Such communications consultants must be experts in both business management and in telecommunications.

□ Telecommunications Manager

Because of the increasing complexities of telecommunications, many large firms employ a specialist at the top management level who is in charge of all the equipment and services devoted to their communications support system, from the station users's instrument to an elaborate worldwide communications network. Such a person may be responsible for millions of dollars in equipment and hundreds of thousands of dollars in monthly service expenditures. Indeed, the knowledge and expertise that must be possessed by the telecommunications manager is a testimony to the scope and importance of the telephone's role in business today.

QUESTIONS FOR REINFORCEMENT AND DISCUSSION

1. Where is the nearest telephone? How would you feel if it rang right now? And rang, and rang, and rang, and rang . . . .
2. Are you usually aware of the telephone in the room? Have you ever been without telephone service for several days? Have you ever been without telephone service at work? In what ways did you find that you generally take the telephone for granted?
3. What good or bad experiences have you had when you called a store or company as a customer or client?
4. What good or bad experiences have you had when you called a government agency or service as a citizen?
5. Can you relate a funny experience you had on the telephone?
6. What is the most exasperating experience you had on the telephone?
7. Has the telephone ever saved your life?
8. Name and describe a public service upon which you rely frequently and tell how the telephone is used to facilitate that service.
9. Name and describe a business service upon which you rely frequently and how the telephone is used to facilitate that service.
10. What is the longest distance over which you have ever communicated by telephone?

SPECIAL PROJECTS

1. Visit a business or government office and spend a few minutes observing the telephone equipment and people who use it. If you can do so without causing an inconvenience, explain that you are studying telephone usage in school and ask several questions you feel are important. Report your observations in writing or orally to the class.
   a. Describe the equipment. Was there a brand name on it?
   b. Do you think special training is required to operate it?
   c. Did it serve the needs of a large or small organization?
   d. Did it appear to serve those needs adequately?
   e. Did the people who used it appear to be prepared for the types of calls they received? For the types of calls they originated?

   f. Did they appear to enjoy their conversations or did they appear to be under pressure?
   g. Do you think the callers got an accurate impression of the office, the company, its service, and its products from their conversations?
   h. What would you have done to improve the company's telephone usage?

2. Choose a local vendor of telephone services or equipment and obtain additional information about it by calling a market representative and requesting a brochure.
   a. Who are the company's customers?
   b. Who are the company's competitors?
   c. Describe the product or service.
   d. How much technical information about it should a person have in order to sell it? How much technical information should be understood by a person who buys it? By an employee who uses it?
   e. How much does it cost? How will it save the buyer money?

3. Schedule an interview with someone who works for a local telephone carrier or vendor. If you cannot visit the person's office or workstation, conduct an interview over the telephone.
   a. Find out the qualifications for the job.
   b. Describe the nature and duties of the job.
   c. Discuss the advantages and disadvantages of the position. The opportunities for promotion.
   d. Describe how the job is related to the telephone.
   e. Ask the person what he or she sees in the future of the telephone and the telecommunications industry.
   f. Ask the person what he or she thinks is the most important thing that should be learned about using the telephone.

Telephone Equipment

As an office worker with a telephone station, you have status, privilege, and responsibility. The instrument on your desk requires skill and knowledge for its operation—more than you may realize. It is not to be taken for granted.

There are many different kinds of telephones in offices today. Some are as simple as the telephone you have at home, and others are actually small computers with many useful features and conveniences.

Let's begin our discussion with the parts that are found on nearly every telephone set. First, there is the *handset*, which includes the *transmitter* (mouthpiece) and *receiver* (earpiece) and is the part you hold in your hand. The other part is the *telephone*, which contains the electrical circuits that perform the various functions of your telephone. There is also a *ringer* (a bell or other audible signal) and the *switch-hook*,[1] which is up when the telephone is in use (off-hook position) and depressed when the telephone is not in use (on-hook position).

The *dial* or *tone buttons* that let you call the number you want deserve special comment. If your telephone set has a dial with a hole in

[1] Also called a hook switch.

it for each number, you call the number you want by sending a series of pulses to the central office of the telephone company. If your instrument has a button for each number, you send a musical tone to the central office. The technical name for tone dialing is Dual-Tone Multi-Frequency (DTMF). Touch-Tone is a registered trademark of AT&T. The primary advantages offered by DTMF are faster and more accurate call placement and compatibility with computers and other business machines.

■ KEY TELEPHONES

First let us consider the most common of the multiline telephones, the key telephone, with its rows of buttons and flashing lights. It is sometimes referred to as the push-button telephone or the Call Director (the registered trade name for AT&T's key telephone). Each button that controls a central office line is called a *key*, which is how this telephone gets its name. It is the workhorse of business telephone systems, and it has probably frightened more inexperienced office workers than any other business machine!

Going from left to right on the six-button instrument shown on page 40, you will see a *hold* button, four incoming *central office lines* (also known as *trunks*), and an *interoffice line* (com6). When a caller dials 555-1674, the ring or audible signal is heard and the light on the first line flashes slowly. As soon as the button for that line is depressed and the switchhook is lifted, the light becomes steady.

If a call is in progress on the first line and another caller dials 555-1674, the call automatically rings in on the second line (555-1675) by means of a rotary, and the light on the second line flashes slowly. The person talking on the first line can ask the first caller to hold, depress the hold button, and then answer the second call by depressing the button for 555-1675. While the caller on the first line is holding, the light on 1674 flashes rapidly as a reminder that someone is still there waiting.

The lights always indicate which line is ringing (slow flash), which line is holding (rapid flash), which line is in use (steady light), and which line is available for use (no light). It is quite easy for one person to handle calls on all four lines and the interoffice line without inconveniencing anyone at any time. The important thing to remember, of course, is to depress the hold button before going to another line so that no one will be disconnected.

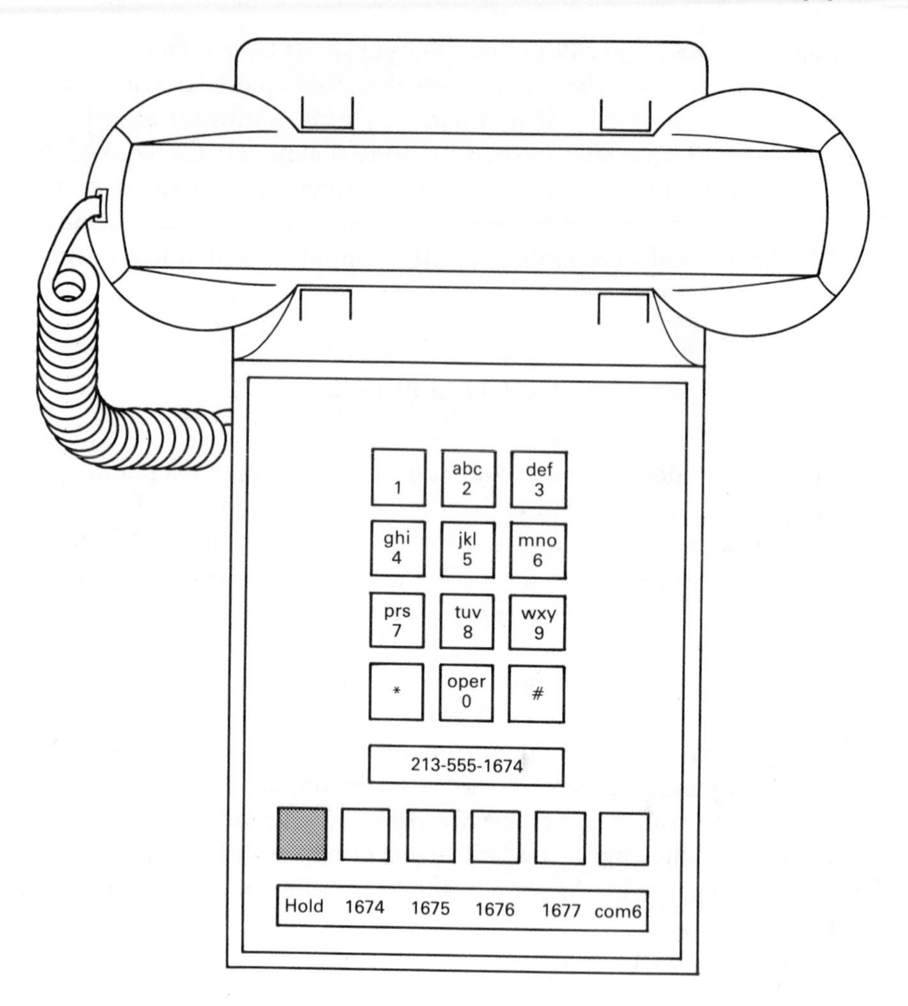

The last button is the interoffice, or intercom, line. It is a closed-circuit system, which ties all the telephones in the office together. If someone in the office on station 6 wants to speak to someone in the office on station 7, he or she can depress the com6 button and depress the number 7 with the handset lifted; station 7 will then ring. When both handsets are off their switchhooks, the line is open and the people can talk. It is not necessary to use the hold button for an interoffice call because there is no connection to break. The "dialing" of the number described above merely rings the bell on the telephone.

Key telephone systems may have from 2 to 20 lines serving as many as 50 stations. Their use is limited to smaller offices, and most have the disadvantage of a lack of privacy—anyone can pick up any line at any time. Some key telephones provide the attendant with busy-lamp signals, which show those stations that are busy (who is talking on the phone and who is available to receive calls). Some even have additional capabilities, such as multiline conferencing (parties at more than two locations may be connected without the assistance of a telephone company operator), music on hold, and "message-waiting" lamps similar to those found in hotel rooms. A few are equipped with Direct Station Selection (DSS), which enables the user to transfer a call to another station by simply pressing a button for the desired station and eliminating the need for the hold button.

If you begin with a good understanding of the key telephone and its operation, it will be easy for you to learn to use the instruments with more-sophisticated features described in the remainder of this chapter.

AT&T's Com-Key™ 2152 business telephone system with 1434 direct station selection. Photo reproduced with permission of AT&T.

■ PRIVATE SWITCHING EXCHANGES

The alternative to a conventional key telephone system is the private switching exchange. The office telephone has often been associated with the term PBX, which means Private Branch Exchange. A company with a PBX actually has its own private switching station, separate from the public switching system of the telephone company, with which it can handle many calls simultaneously on its own published numbers. Originally, a trained PBX operator (or team of operators) manipulated cords and plugs to achieve functions such as routing (switching calls to the proper station), transferring (switching calls from station to station), holding calls for busy stations, reminding stations that calls are waiting, placing calls again if the lines are busy, and helping to control telephone costs.

Although cord-and-plug PBX switchboards are still in use, they have for the most part been replaced by automatic switching equipment, usually called PABXs (Private Automatic Branch Exchanges), which can accomplish many or all of these tasks. Electromechanical switches have been used by businesses for many years and are sometimes called "dumb" switches, when compared to the electronic switches now appearing. Electronic switches are sometimes called "smart," or "intelligent," switches because they are actually computers that offer many useful features (discussed later in this chapter). Other terms for electronic switches are EAPBX (Electronic Automatic Private Branch Exchange) and CBX (Computerized Branch Exchange). In the final analysis, manual switchboards, electromechanical switches, and electronic switches are all private branch exchanges, and all may be called PBXs.

Both key telephones, such as the one described earlier, and single-line telephones may be used with PABXs. With a single-line set, the caller may be placed on hold by a quick depression of the switchhook, and an outside line may be obtained by dialing an access code—normally a 9. Many of today's electronic switches offer users both Direct Inward Dialing (DID) and Direct Outward Dialing (DOD), which means neither an access code nor a company operator is required. More is said later about the operation of telephone sets connected to modern switches.

■ CENTREX

CENTREX is actually a private electronic switching exchange maintained by the telephone company for the business user. It offers many advantages to those companies that need very large systems but do not want to administer them. Switching equipment is commonly located at the central office of the telephone company, and each station in the customer's company has a separate line or telephone number. A block of numbers with the same prefix (first three digits) is assigned to the company. Both key sets and single-line instruments may be used with CENTREX, and users have both DID and DOD. Interoffice calls are made by omitting the prefix and dialing only the last four digits of the station number.

■ ELECTRONIC KEY TELEPHONE SYSTEMS

The electronic key telephone system is like the key telephone because it is operated primarily by pushbuttons or keys. However, it is part of an electronic switching system that is controlled by a microprocessor, which enables it to perform many functions that conventional key telephones cannot do. In fact, electronic key telephone systems can be designed by the equipment manufacturer to suit the special needs of the customer. They can then be programmed and reprogrammed by the customer to meet the changing requirements of its organization. Stations can be moved, added to, or dropped from the system, and some features can readily be changed or adjusted. This is a tremendous advantage in today's changing business world.

Electronic key telephone systems range in size from 12 outside lines serving 20 stations up to 20 outside lines serving 100 or more stations. Some systems may have several intercom paths. Features offered may be classified as (1) those that assist the attendant or operator and facilitiate the use of the intercom, (2) those that assist the user and facilitate communication among people, (3) those that facilitate call placement, and (4) those that facilitate cost control.

☐ Features That Assist the Attendant and Facilitate the Use of the Intercom

Call Monitoring. One of the most valuable features of the electronic-key telephone is the monitor, which is a built-in speaker for the intercom. When a call is routed to a station, the person at that station hears an audible signal and can immediately talk to the attendant through the monitor without lifting the handset or depressing a button for the intercom line. This is sometimes called *hands-free*, or *on-hook, answerback*.

Direct Station Selection with Busy-Lamp Field (DSS/BLF). A telset equipped with DSS/BLF has a bank of buttons with one button for each station in the system. When a station is busy, the button for that station lights up. In addition, the attendant (or someone using the intercom) can route the call by simply depressing the button for that station.

Call Waiting Signal. With the call waiting feature, the attendant can use a soft tone to signal a busy station that a call is waiting; only the station can hear the tone (the outside caller cannot hear it). A call-waiting signal can also be made with a low-volume announcement (soft page) through the intercom monitor.

Call Forwarding. If a user is going to be away from his or her desk, the unattended telephone station can be programmed so that calls are automatically routed to another station in the system at which the user can be reached or at which someone can take the calls or take messages.

Paging. A staff member may be paged through the intercom path on the monitor. Some systems are programmed for *zone paging*; with this, the attendant can page a specific group of stations, such as an entire department.

Message Waiting. The attendant can signal that a message is waiting by activating a light on the station user's telset. Some electronic key telephone sets even have small visual displays that can show a very brief written message.

Override. The attendant can break in on a call in progress for an important incoming call or emergency. This is sometimes referred to as *executive override* because the system can also be programmed to enable an executive or supervisor to break into calls of subordinates.

Attendant Callback. The attendant never needs to worry about forgetting a caller on hold. The system can be programmed so that the audible signal is heard again after a specified number of seconds to remind the attendant that someone is still waiting.

Night Transfer. When the attendant goes home at night, the system is reprogrammed so that incoming calls are automatically forwarded to a department operating a night shift, to a security guard station, or to any station that wants to take calls after regular business hours.

☐ Features That Assist Users and Facilitate Communication Among People

Privacy. Unlike the conventional key telephone, every outside and intercom connection is completely private.

Do Not Disturb. The station user can program the telset so that the call-waiting signals cannot get through and so that the override cannot be exercised.

Direct Conferencing. The user can arrange conferences without the assistance of the attendant or the telephone company operator. This means that more than two parties from both inside and outside the company can be connected for one call. Conferences can be arranged on outside lines, as well as on the intercom paths.

Distinctive Rings. The audible signal (ring, tone, or beep) announcing a call can vary to indicate the nature of the caller. For example, the signal for a call coming from outside the company may be different from the signal for a call on the intercom. A special signal may be used for an executive or supervisor calling subordinates or for calls between an executive and a secretary.

Station Callback. As with the attendant callback, the station user never needs to worry about leaving a caller on hold. The system will sound an audible signal after a call has been on hold for a specified number of seconds.

Hands-Free (On-Hook) Intercom. Station users can carry on two-way conversations or conference calls on the intercom paths without lifting their handsets or receivers.

☐ Features That Facilitate Placing Outside Calls

Automatic Dialing. The electronic key telephone set has a memory. This means it can be programmed to remember certain frequently called numbers (as many as 32 in some cases). The station user can then call those numbers with a simple two- or three-digit access code. This features is sometimes called *speed calling.*

Last-Number Redial. The electronic key telephone set also remembers the last number dialed. When a call reaches a busy line or is not answered, the station user can place the call again automatically with a two- or three-digit access code. Significant time can be saved when it is not necessary to dial the entire number again.

Automatic Recall on Intercom. If a call is placed on an intercom path to a station that is busy, the caller can remain on-hook, and the system will signal when the busy station is free so that the call can be placed again.

Hands-Free (On-Hook) Dialing. The receiver does not have to be lifted when calls are dialed either on the intercom paths or on the outside lines. The dial tone, the ringing (or busy signal), and the answer can be heard through the monitor. When the call is completed, the user can lift the handset, and the call becomes private.

□ Features That Help Control Telephone Costs

The class of service available to certain stations can be limited. This means that the telset is programmed so that its class of service includes only intercom calls, only local calls, or only specified long-distance calls. If a call is dialed to a destination outside the restricted area, the caller will get a busy signal or a special tone to indicate class-of-service limitation. This arrangement is also called *toll restriction.*

■ COMPUTERIZED TELEPHONE SYSTEMS

The main difference between an electronic key telephone system and a computerized telephone system is in the capacity of the computer, rather than in the technology or operation. Large electronic switching

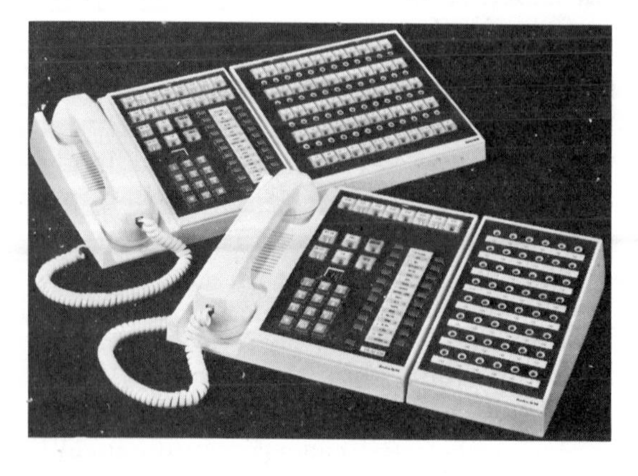

NEC's Electra 16/48 key telephone systems, in which each key phone contains its own microprocessor. Photograph courtesy of NEC Telephones, Inc.

systems can handle hundreds of trunks and thousands of lines. Their increased capacities permit them to offer all the features of the electronic-key telephone systems described earlier, plus the even more sophisticated features described next.

Hunting. Incoming calls automatically hunt through the stations in a designated group until one that is not busy is found. Most hunting is consecutive—that is, the call proceeds through the stations starting with the highest number and working down. With consecutive hunting, however, the stations with the higher numbers tend to get the most calls. Circular hunting distributes calls more evenly by seeking the next number down from the last number called. In this way, the station that just received a call will not get another call until all the stations in the group have been tried.

Camping. When an incoming call encounters a busy line, it is automatically placed in a "camp-on" mode until that station is free. The call then rings through automatically without attendant assistance.

Queuing. Calls that are camping wait in line and are automatically put through on a first-come—first-serve basis, just as if the customers were standing in line at the counter.

Automatic Call Distribution (ACD). Automatic call distribution is a combination of hunting, camping, and queuing in which calls are distributed in an organized manner among a designated group of stations. Music or a recorded message may be played while the calls are in a waiting mode.

Consultation on Hold. Even a station with a single-line telephone can place a call on hold, call either an inside party or an outside party for consultation or to get information, and then return to the call on hold to continue the original conversation.

Call Pickup. A user can pick up calls for a station other than his or her own by dialing an access code.

Call Park. A user can hold a call on his or her station, go to another office—perhaps to get some information or a file—and retrieve the call on another phone.

More Audible Signals. In addition to the audible signals that indicate where the call may be coming from, computerized telephones provide signals assisting the user in the operation of the equipment.

Two such audible signals with which you are already familiar are the dial tone and the busy signal. You also learned about the special tone that may be used for toll restriction. Other operational signals may include a confirmation tone, indicating that an operation has been accepted by the system; an error tone (sometimes called a howler), indicating that the operation has gone wrong and must be started again; and a recall tone, indicating that a call waiting for a station to be free can now be placed again.

Least-Cost Routing (LCR). The computerized telephone system can be programmed to choose the line or class of service that is the least expensive. For example, if the company is a WATS subscriber (WATS service is discussed in the next chapter) and a user dials a long-distance call, the call will automatically be placed on the WATS line rather than on a local line. For companies with elaborate networking systems, such as those described in Chapter 5, this capability results in substantial savings. It is sometimes called *route optimization,* or *automatic route selection (ARS).*

Mitel's SX-5 Superswitch is a compact, but powerful, computerized switching system. Reproduced with permission of Mitel, Inc.

The model in this picture is pointing to the 32-bit computer shelf in one *node* (cabinet) of the multinode ROLM® VLCBX. The VLCBX is capable of handling up to 4000 telephone extensions. Photo courtesy of ROLM Corporation.

Call-Detail Recording (CDR). In addition to assisting in the routing and placing of calls, the computer system also creates a record of all calls, when they were placed, from what station they were placed, on what line they were placed, their destinations, and how long they lasted. The system can then periodically analyze these data to pinpoint possible abuse and to help the company find ways to reduce its telephone costs. More about these systems is presented in the chapter on costs and controls.

Traffic Analysis. Traffic data—such as how many calls come in and go out, frequently called locations and peak usage times—are recorded. These data are then analyzed to help the company use its system more efficiently. For example, if too many incoming calls are encountering busy signals, the company may be losing business and want to add more trunks.

Directory Look-up System (DLS). The storage capacity of the computer can be used to maintain a complete directory of all the station users in the system, as well as other information desired. It may include names, titles, functions, office locations, employee badge numbers, and—of course—telephone numbers. CRT (Cathode Ray Tube) screens can be provided for console operators, receptionists, security guards, or any other personnel who need to locate company employees. Finally, a printer can be linked to the system so that hard copy of the directory can be produced for distribution.

System Diagnostics. The system can actually test itself and report problems that may require maintenance and repair service.

Interface with Other Business Machines and Systems. Computerized telephone systems can communicate with other business machines through connections known as *ports.* Such communication may be desired with other computers, word processors, photocopiers, dictation equipment, and data communication systems. More about these systems is presented in Chapter 5.

Components of the Anderson Jacobsen Integrated Office Exchange (AJIOX) include printers, visual display terminals, the Digi-Touch Telset,(TM) and the IOX computerized switch. Photograph courtesy of Anderson Jacobsen, Inc.

■ EQUIPMENT OPERATION

It is not possible to describe here the operation of all the equipment used with electronic key systems and computerized systems. Most systems have three distinct components: the computer, or processing unit; the attendant, or console operator station; and the user stations. User stations may be either single-line or electronic key telephones. The processing unit is generally programmed and operated by trained specialists, although some of the operations may be reprogrammed and changed by the attendant. Attendant training is often provided by the manufacturer of the equipment.

Some of the user operations are the same as those performed by the attendant. Your company will provide you with any special training or instruction manuals that are necessary, or the manufacturer of the equipment may train your entire staff when the equipment is installed. Generally speaking, operations are performed with one of the following four basic functions.

1. Flashing. Some operations are performed by a quick depression of the switch-hook. This is called *flashing* because it flashes a signal to the system or to the attendant. On AT&T's Dimension system, a call is placed on hold by flashing. Some newer single-line telephones have a separate button for performing the flashing function.

2. Function keys. The most familiar function key is the hold button. When it is depressed, the telephone is instructed to place the call in a waiting mode. Another function key may be used to create a conference call, such as the ADON key on GTE's EKTS electronic key telephone set, which instructs the telephone to add another connection on the line. On the electronic key set used with Northern Telecom's SL-1, a key marked CFW actuates the call-forward feature.

3. Access codes. The most common access code is the number 9, dialed to get an outside line. Similarly, some systems access a long-distance line with the number 8.

4. Starring and Pounding. The * and # symbols on touch-dialing systems, such as Touch-Tone, may be combined with digits to create instructions. The * may be called either the *star* or the *asterisk*, and the # may be called either the *pound* or the *sharp* sign. Starring and pounding are similar to function keys in that they actuate certain features. For example, in GTE's computerized system, the automatic dialing feature is actuated with the # (pound) plus 6 plus a code number.

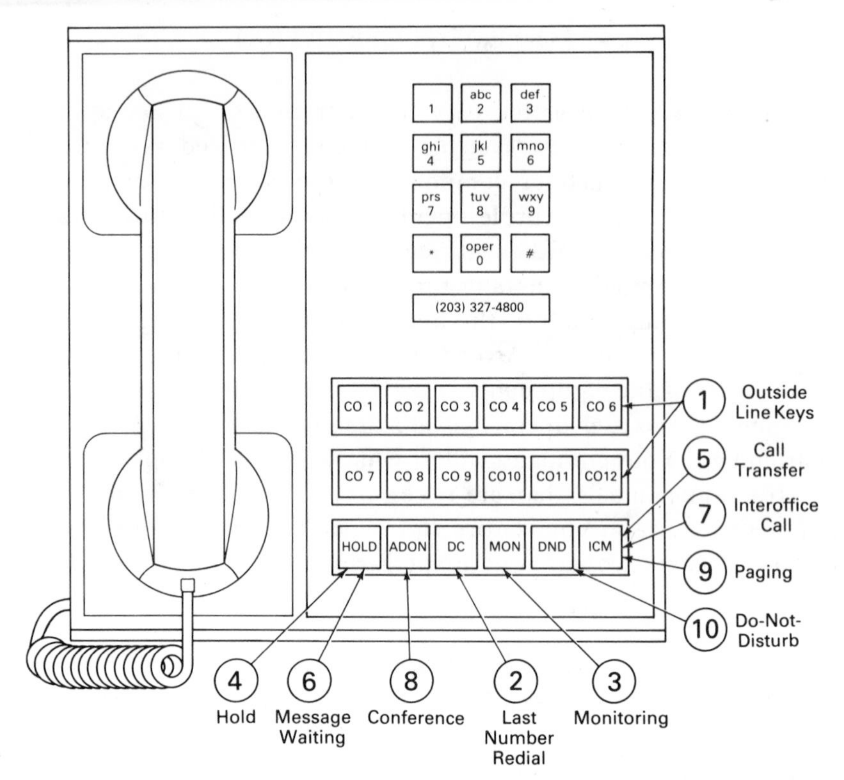

This illustration of GTE's EKTS electronic key telephone set shows 12 incoming lines and function keys. Reproduced with permission of GTE Telecommunications Center.

■ ADDITIONAL EQUIPMENT

Some of the features integrated into computerized systems can be acquired separately and operated in conjunction with conventional telephone equipment. In addition, there are a number of auxiliary devices and equipment systems of which you should be aware.

□ Hands-Free Devices

Office workers are frequently required to write or perform other tasks while talking on the telephone. Console attendants must also have their hands free while they are answering and routing calls or taking messages. For this purpose, comfortable, lightweight headsets are available in styles to suit the personal preferences of the users. For those who do not want to wear headsets, shoulder-rest attachments may be purchased.

The most effective hands-free device, however, is a special telephone set, such as AT&T's Speakerphone® or conference telephone. Such a telephone actually uses an amplifier, in addition to the microphone and transmitter of the telephone, to broadcast a telephone conversation. The user (and everyone else present) can move around the room to find files or continue working while talking on the telephone. Portable conference telephone sets that can be carried as conveniently as briefcases and plugged into electrical outlets with AC adapters are available. Special hands-free telephone sets are also available for the handicapped.

☐ Picturephone®

As its name implies, AT&T's Picturephone® allows users to see each other as they talk. They can also see documents, models, and other objects under discussion. Videoconferencing services are discussed in the next chapter.

☐ Dialing Devices

Separate automatic dialing devices can be purchased for use with conventional telephones for speed and convenience similar to those described on the electronic key telephone. Elaborate variations enable the user simply to push a button, speak into a monitor, and continue working while the call is being dialed automatically, routed through an operator, or placed on hold. When the person called comes on the line, the user picks up the receiver and the call becomes private. Some dialers can store and dial any number of digits for several hundred names and numbers. Numbers are easily entered and changed by the user.

☐ Telephone-Answering Equipment

Nearly three-quarters of all telephone calls placed are not completed. Either the line is busy or the person called is not available. Today's businessperson just cannot be tied to the telephone. Nor can the game dubbed *telephone tag* by *Business Week* magazine be tolerated. (Telephone tag is a game in which two business persons leave messages for each other all day long. The last one to answer the message tries to remember what the call was about!)

The Zelex 910 electronic telephone directory displays the complete name and number of the person being called and places the call automatically. It stores up to 340 names and numbers. Photo courtesy of Zelex Corporation.

Some executives have constant telephone coverage provided by their secretaries, but for most office workers, such coverage is an unaffordable luxury. Many office workers cover phones for each other, but these arrangements become most inconvenient. Automatic telephone-answering devices and systems can relieve the problem.

Telephone-answering devices can be as simple as a tape recorder attached to a telephone. The device answers the phone with a recorded message and the caller leaves a message on the cassette tape. The user then listens to the tape upon returning to the office. More elaborate answering equipment offers the following additional capabilities:

1. The message recording device can be accessed from any telephone anywhere in the world. The user calls his or her number and signals the device to play out the messages over the phone. The recorded message can even be changed and rerecorded from a remote telephone.
2. The answering device can be used as a screening device. The user can listen as the call is recorded and decide whether to answer immediately while the caller is on the line or return the call later.
3. Some answering equipment includes a call-diverting feature that will forward the call to another number or hunt through a series of preprogrammed numbers until the user is found. The device literally "tracks down its master" with important calls or emergencies.

The Code-A-Phone 1750 offers voice-controlled message capacity, plus remote command. Photo courtesy of Ford Industries, Inc.

4 Messages as long as 30 minutes can be recorded. Elaborate instructions can be transmitted, and whole orders can be recorded by customers.

A firm with a large sales force or team of field consultants can maintain a centralized message-recording system with capabilities such as those described above to enable employees to keep in touch with callers 24 hours a day from any remote location in the world without regard to time zones or office hours.

☐ Voice Messaging Equipment

Voice messaging may best be described as a paperless, interoffice mail system. Instead of writing memos, users send messages by telephone. These messages are stored in a computer, which notifies recipients that they are waiting. The receiver can request and listen to messages from any telephone. Messages can then be either stored permanently or deleted. The resulting savings in paper and clerical costs easily justify the cost of such equipment for larger organizations.

☐ Automatic Call Distribution Systems

Customers will not wait. When a customer gets a busy signal (known as a *blocked call*) or becomes tired of waiting on hold (called an *abandoned call*), the customer hangs up. Blocked and abandoned calls result in lost business and are therefore very expensive. ACD equipment can be purchased and installed for use in conjunction with conventional

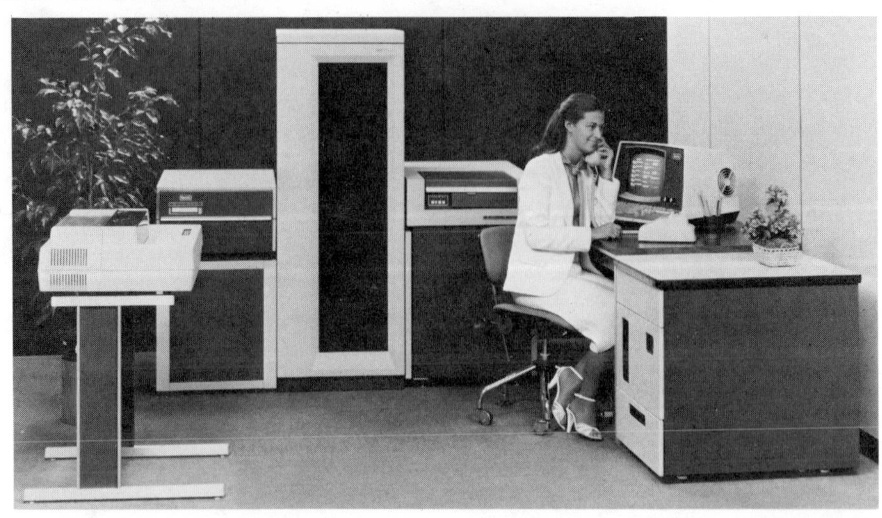

Wang Laboratories developed this Digital Voice Exchange (DVX) in response to research showing that managers strongly prefer verbal communications. Photo courtesy of Wang Laboratories.

equipment to perform the same functions as the ACD feature integrated into the computerized system described earlier in this chapter. Incoming calls are automatically routed to the next available line so that faster, more convenient service is provided to customers. A recorded message informs the caller that someone will be available soon, and music-on-hold makes the waiting easier.

□ Paging Systems

Paging systems enable employees such as supervisors, repairpeople, and field representatives to remain accessible even though they are required to move around a great deal both on and off the company premises. Such paging systems may be either separate or integrated into the telephone system. An access code enables users to place their stations in the paging mode and broadcast messages to groups of stations on the intercom system or through strategically located loudspeakers on the premises. Pocket-sized, remote receivers—sometimes called *beepers*—may be carried off the premises so that field personnel may be reached instantly wherever they may be within a broad geographical area.

□ Mobile Telephones

There are several types of telephones that can be moved about. One is the mobile phone, which can be installed in cars, trucks, planes, and boats. Communications travel both by radio and by telephone lines. Another type is the portable telephone sets, which can be plugged either into a telephone jack or into a wall outlet by means of an AC adapter. Fort short-range mobility, there are cordless, battery-operated sets that can place and receive calls anywhere within several hundred feet of a base telephone. A special transmitter is attached to the base telephone and the cordless set must be recharged after a certain period of time.

□ Telecopiers

Photocopiers have been able to "talk" on the telephone for a number of years. A telecopier at one location can photograph a document and send it by telephone to a telecopier in another location, which reproduces that document on paper. The process is known as *facsimile* (abbreviated FAX) and is considered to be a form of electronic mail.

The cordless handset of this telephone can be carried up to 600 feet from its base set. Photo courtesy of Extend-A-Phone.

One important advantage offered by FAX is that drawings, maps, charts, and handwritten documents as well as printed text, can be transmitted. Technological advancements are bringing the cost of FAX down by reducing the transmission time to under 1 minute for each document, which means lower telephone charges. Machines that operate unattended and feed documents automatically enable companies to schedule transmission at night, in order to take advantage of lower telephone rates. Portable telecopiers are also available.

☐ Teletypewriters

Teletypewriters and teleprinters are terminal equipment used for transmission of written messages over telephone lines. Teleprinters are used in conjunction with data communications sytems and record communications services such as those described in the next chapter. They are also used to enable the deaf to communicate over the telephone.

The operation of these auxiliary devices is usually very simple and requires little or no training. However, in order to make the best use of all the equipment described in this chapter, you also need to know more about the services offered by the many carriers whose

Siemens T-1000 electronic teleprinter. Reproduced with permission of Siemens Corporation.

facilities transmit our telecommunications all over the world. These are discussed in the next chapter.

QUESTIONS FOR REINFORCEMENT AND DISCUSSION

1. Why do most businesses and offices have multiline telephone systems?
2. What does the hold button do on a conventional key telephone? Which line does not require the use of the hold button? Why?
3. When might it be necessary to place a caller on hold?
4. What is the meaning of each of the following?
   a. Steady light
   b. Rapidly flashing light
   c. Slowly flashing light
   d. No light
5. Describe DTMF and state its two principal advantages.
6. Describe the following features on an electronic key telephone.
   a. Call monitoring
   b. Zone paging
   c. Last number redial
   d. Night transfer
   e. Class of service
   f. Soft page
7. Tell how you think the following features would help an office worker.
   a. Call-waiting signal
   b. Station callback
   c. On-hook dialing
   d. Call forwarding
   e. Distinctive rings
   f. Automatic redial on intercom
8. What is toll restriction?
9. What are some ways an electronic key system can be reprogrammed to meet the changing needs of an organization?
10. What is an abandoned call? What is a blocked call? Why are they important in business?
11. List some of the capabilities of telephone message-recording devices.
12. Describe each of the four operations for single-line and electronic key telephone sets that are used with computerized telephone systems.
    a. Flashing
    b. Function keys
    c. Access codes
    d. Starring and pounding

MATCHING QUESTIONS (PART A)

Match each of the following items to its definition by writing the correct letter on the blank provided. You should use every letter once. Refer to your book as needed. This exercise is intended to help you learn, not to test you.

a. Abandoned call
b. Beeper
c. Blocked call
d. CENTREX
e. Com line
f. Conferencing
g. Cordless phone
h. Handset
i. Hold button
j. Mobile phone
k. Picturephone
l. Portable phone
m. Pounding
n. Speakerphone
o. Smart switch
p. Starring
q. Switch
r. Telephone tag
s. Telset
t. Videoconferencing

_______ 1. Telephone equipment on which you can see and talk at the same time.
_______ 2. Leaving messages for each other all day.
_______ 3. Mobile phone that can be plugged in anywhere.
_______ 4. Mobile phone that operates within a certain range of a base phone.
_______ 5. Computerized switching system maintained by the telephone company.
_______ 6. Electronic or computerized switch.
_______ 7. Combination of telephone and television for a meeting.
_______ 8. Nickname for telephone instrument.
_______ 9. Call that encounters a busy signal.
_______ 10. Hands-free phone, also called a conference telephone.
_______ 11. Has a transmitter (mouthpiece) and a receiver (earpiece).
_______ 12. Nickname for the receiver used in a remote paging system.
_______ 13. Interoffice phone line.
_______ 14. Caller gets tired of waiting and hangs up.
_______ 15. Equipment that connects telephone callers.
_______ 16. Instructing a computerized telephone by pressing #.
_______ 17. Phone using both radio and telephone transmission from a car.
_______ 18. Function key that allows a call to be placed on hold.
_______ 19. More than two parties involved in one call.
_______ 20. Instructing a computerized telephone by pressing *.

MATCHING QUESTIONS (PART B)

Match each of the following features to its definition by writing the correct letter on the blank provided. You should use every letter once. Refer to your book as needed. This exercise is intended to help you learn, not to test you.

a. Attendant callback
b. Automatic dialing
c. Automatic recall
d. Automatic redial
e. Busy lamp field
f. Call forwarding
g. Call park
h. Call pickup
i. Call-waiting signal
j. Camping
k. Direct station selection
l. Howler
m. Hunting
n. Monitor
o. Night transfer
p. Override
q. Queuing
r. Station callback
s. System diagnostics
t. Zone paging

_______ 1. Holding calls in line.
_______ 2. Audible signal that an instruction is not accepted by the system.
_______ 3. Routing calls by pushing a single button, without using the hold button.
_______ 4. Audible signal reminding the attendant that a call is waiting.
_______ 5. Audible signal reminding the station user that a call is waiting.
_______ 6. Speaker on electronic-key telset allowing hands-free intercom use.
_______ 7. Programming to page a specific group of stations.
_______ 8. Locating and reporting trouble in the system.
_______ 9. Holding for a busy station and automatically ringing when it is free.
_______ 10. Answering a phone at a station other than the one dialed.
_______ 11. Low tone indicating a call waiting, heard only by the person called.
_______ 12. Placing a call on hold so it can be picked up later at another station.
_______ 13. Signaling when a busy intercom line is free and can be tried again.
_______ 14. Field of lights telling which stations are busy.
_______ 15. Programming so calls are automatically forwarded to another station.
_______ 16. Numbers stored in memory and dialed automatically.
_______ 17. Last number dialed stored in memory and dialed again.
_______ 18. Call automatically searching for a station that is not busy.
_______ 19. Programming incoming calls to ring at certain stations after hours.
_______ 20. Breaking in on a call.

Services

■ LONG-DISTANCE VOICE COMMUNICATIONS

If you have always thought of a long-distance call as an expensive luxury, you must now change your way of thinking. Long-distance communication is probably the most important of all the services offered to telephone users, especially in business. Long-distance communication, in fact, represents daily activities that are important contributions to the profits of most companies. Long-distance voice communication is an expensive necessity.

Consider the value of the long-distance call that provides expedient service to a customer requiring information or products that must be obtained from a distant branch. Consider the importance of daily communications among branch-operation personnel. Consider the competitive advantage of instant market information from far corners of the world. Consider the growth potential in a service or marketing area that can be expanded by the telephone rather than by travel or opening a branch office. These are business advantages made possible by long-distance voice communications that we take for granted.

Finally, consider the cost savings of a telephone call over a letter: the cost of a typed business letter is now estimated to exceed $5, while a call can be made from Los Angeles to New York City for under $1 during reduced-rate periods. Certainly the telephone will never replace letters, traveling, or branch offices, but it is an attractive and well-established alternative.

□ Dialing a Long-Distance Call

Station-to-station Direct Distance Dialing (DDD) is the cheapest, fastest, long-distance service offered on the public networks. Requiring no operator assistance, a connection can be made by dialing eleven digits, which have the following functions.

Access to DDD	*Area Code*	*Exchange Prefix*	*Station Number*
1	213	555	1674

Dialing 1 for access to the direct distance dialing network of AT&T's Long Lines Department is necessary in most areas. The area code refers to the telephone area in the United States or Canada, the prefix dials into the local exchange office, and the last four digits reach the station of the business or residence. If you do not know the prefix and station number of the business or residence you want to call in another area code, you can obtain it by dialing the long-distance access number and area code, followed by 555-1212. There is no charge for this long-distance information service. If you do not know the area code, you can consult the white pages of your telephone directory or call the directory-assistance operator in your local exchange.

If you make an error in dialing, dial an incorrect number, or get a bad connection, you can have the operator cancel the charges by immediately dialing 0 and reporting the error. The operator will then place the call again for you or ask you to place it again yourself.

□ International Direct Distance Dialing (IDDD)

Many foreign countries can also be dialed directly at relatively low costs. This dialing service requires an international long-distance access number 011 instead of 1, followed by a country code and the telephone number. A complete listing of country codes may be found in the white pages of the telephone directory. Rates for international long-distance calls made from the United States are lower than any other country in the world.

☐ Operator-Assisted, Long-Distance Calls

Some long-distance calls do require operator assistance, and the service provided by the operator is well worth the small additional cost. In areas where the access number 1 is used, operator-assisted calls may be placed by dialing 0, followed by the area code, prefix, and station number. The operator will then come on the line before the call is completed. In areas where the access number is not used, the operator will dial the number for you. Long-distance operator services include the following.

Person-to-Person. This is the most expensive type of long-distance call. The operator will not leave the line until the person you want has answered the phone, and no charge is made unless the call is completed.

Credit Card Calls. Credit cards afford users the convenience of having calls charged to a specified monthly bill from any telephone. In addition, the bill will show charges to specific credit cards so that companies can allocate costs to certain departments or individuals.

Third-Number Calls. Persons who do not have credit cards can have long-distance calls charged to an authorized third number. Such service might be used when making a call from a customer's phone or when making personal calls from an employer's phone.

Collect Calls. Charges may be reversed if the party called agrees to pay. The operator will leave the line only after the call has been accepted. In business, some companies will accept collect calls from customers. As an employee, however, you should not accept such a call unless specifically authorized to do so by company policy or by your supervisor.

Time and Charges. If you want to record the time and cost of the call immediately, you can ask the operator to report that information back to you after the call has been completed. This service is particularly useful for attorneys, consultants, and hotels, who must charge expenses back to clients or guests.

Conference Calls. The long-distance operator can arrange a connection among people in several places at once with conventional telephone equipment, as well as with electronic telephone equipment.

Mobile and Marine Calls. Many calls to mobile phones must be made with operator assistance.

International Calls. Calls to some foreign countries still require operator assistance.

☐ Time Considerations

Before you place a long-distance call, you should be aware of what time it is for two reasons: (1) Lower rates are offered during nonbusiness hours, and (2) the time will be different at your destination if you are calling to another time zone.

Rate Periods. The rate period is the time of day that a call is placed. The three rate periods used by Bell companies and independents using AT&T Long Lines are as follows:

Weekday:	Monday through Friday from 8 A.M. to 5 P.M.
Evening:	Sunday through Friday from 5 P.M. to 11 P.M.
Night/ Weekend:	Every day from 11 P.M. to 8 A.M. and from 8 A.M. on Saturday to 5 P.M. on Sunday.

For all interstate calls, the rates are 35 percent lower than weekday rates during the evening rate period and 60 percent lower during the night/weekend period. (Reduced rates for calls made within some states may vary.) Because of these significant rate reductions, some companies maintain special shifts so that employees can make long-distance calls during advantageous rate periods.

Time Zones. When you place a long-distance call to a person in another time zone, you want to be sure first that the office is open and second that the person you are calling is available. In this respect, time zones create *windows of communication*. People are generally available between 9 and 11 A.M. and between 2 P.M. and 5 P.M. Times between 8 A.M. and 9 A.M., noon and 2 P.M., and 5 P.M. and 6 P.M. might be risky.

Ideally, when you are calling from one time zone to another, you want the windows to be open at both your location and at your destination; that is, you want the call to occur during regular business hours in both time zones. You can, of course, open the windows at your location if you are willing to arrive at work early, give up your lunch, or stay late. However, you cannot open the windows at your destination unless you are willing to ask the person you call to make those sacrifices.

You can use the time zone chart to determine when the windows are open or closed and when the time is risky for a long-distance call from one time zone to another.

To use the chart, follow the horizontal lines, which show the time variation from one zone to another. When it is 4 A.M. in the Pacific Zone, it is 5 A.M. in the Mountain Zone and 7 A.M. in the Eastern Zone. Note that vertical areas outline the entire time zone. They do not tell time. If it is 4 A.M. in Seattle, it is 4 A.M. in Los Angeles or any other location in the Pacific Zone.

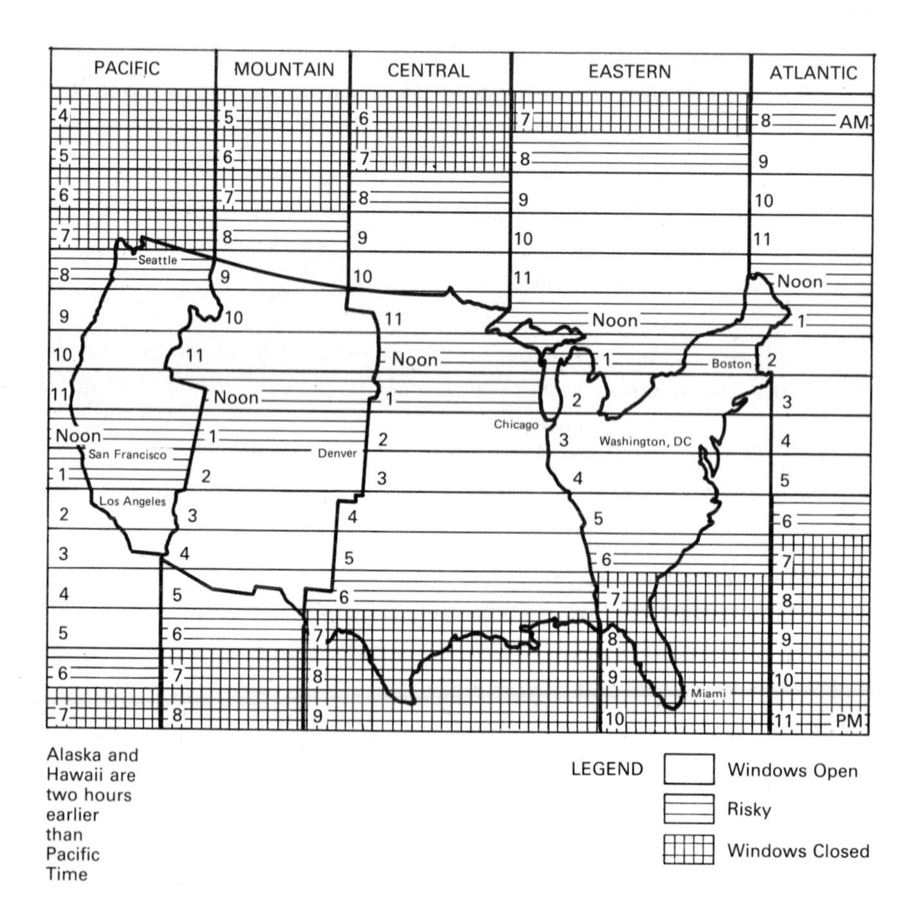

Now see if you can follow these examples.

1 If you are in Washington, D.C., at 10 A.M. and you want to call Los Angeles, it is 7 A.M. in Los Angeles (and everywhere in the Pacific Zone), and the windows are closed.
2 If you are in Denver at 2 P.M. and you want to call Miami, it is 4 P.M. in Miami and the windows are open.
3 If you are in Seattle at 10 A.M. and you are calling Chicago, it is noon in Chicago and it is risky.
4 If you want to call someone in Boston when it is 2 P.M. and you are in Denver, you will have to call at noon Denver time.
5 The only time the windows are open and there is no risk in calling from San Francisco to Miami is between 1 P.M. and 2 P.M. in San Francisco and 4 P.M. and 5 P.M. in Miami.

Making international calls is further complicated when crossing the international dateline; for instance, when it is noon in New York, it is 1 A.M. *the next day* in Manila.

One final note: Be sure you allow for Daylight Savings Time during the appropriate months of the year. It is followed by all states except Arizona, Hawaii, parts of Indiana, Puerto Rico, Virgin Islands, and American Samoa.

☐ Wide Area Telephone Service (WATS)

WATS is a special long-distance service offered by telephone companies using AT&T's Long Lines to users with large volumes of long-distance calls to certain areas. Although WATS can reduce long-distance expenses when properly used, it represents a significant cost to its subscribers. Because many subscribers advertise their WATS lines as "toll-free" to customers and because WATS lines used to be charged on a flat rate, many people have the mistaken idea that WATS calls are always free. Nothing could be further from the truth. Here are some important facts that office workers should know about WATS.[1]

1 WATS is not free. The call is paid for by one of the parties, just like any other long-distance call. Measured-time charges are applied to WATS lines according to hours of usage.
2 WATS calls travel on the same telephone lines as any other long-distance calls. The advantage offered by the service is in the pricing arrangement, not in the method of service.

[1] Larry A. Arredondo, *Telecommunications Management for Business and Government*, 2nd ed. (New York: The Telecom Library, Inc., 1980), pp. 83–85.

3 It is actually possible for an intrastate WATS call to be more expensive than a direct-dialed call placed during a reduced-rate period.
4 A WATS lines goes only one way. WATS lines that receive calls are called IN-WATS, and WATS lines that send calls are called OUT-WATS. The area code for IN-WATS lines is always 800. Rates for IN-WATS are not the same as those for OUT-WATS.
5 A WATS line can be either interstate (from state to state) or intrastate (all calls within one state), but it cannot be both. This accounts for the advertisements you see giving a toll-free 800 number with a separate number for callers in one specific state.
6 WATS lines are effective in certain areas only. The United States is divided into six WATS bands (also called service areas, or SAs), allowing subscribers to reach only the bands desired. Rates are based on distances called *rate steps*, as well as on the amount of use.

☐ Message Units

Charges for most local calls are included in the flat monthly rate for telephone service. Local calls are those calls made within the area served by the local telephone exchange office. Charges to nearby exchanges are, like long-distance calls, measured—this means they are based on the distance in miles and the length of the call in time. The unit of measure for this time and distance is usually called a Message Unit (MU). Message unit computations vary from telephone company to telephone company and cannot be covered in detail here. Measured charges for local calls are already in use in some areas.

If you live in an area where you make calls to many different exchange prefixes, you need to be informed about which prefixes are in your local exchange area. Some firms have so many calls to nearby exchanges that they subscribe to a service called Extended Area Service (EAS), which permits calling into nearby exchanges at reduced rates. For more information about the prefixes included in your local service area and how message unit charges are applied, you should consult the white pages of your telephone directory.

☐ Foreign Exchange Service (FX)

All the telephone company exchanges outside your local service area are known in telephone jargon as "foreign" exchanges. In large metropolitan areas, foreign exchanges may not be so far away—in fact, the customer across the street may be a foreign exchange! If your company has a great deal of traffic to a nearby foreign exchange (or even to a distant exchange across the city or in another state) you may have a direct line to that exchange in order to save on your telephone bills. Such a line may be called an FX line.

Such direct lines mean that you will have lines with different prefixes on your push-button telephone set. You must be sure you make local calls on your local line and applicable calls to foreign exchanges on your FX line. In order to do this, you must know the specific geographic locations of those prefixes you dial frequently and which line on your telephone should be used for them. You can get this information from the white pages of your telephone directory. If you call the wrong prefix on the wrong line, you will be making an unnecessary toll call. It would be as if you drove from Pittsburgh to Baltimore by way of Miami!

□ Private Lines

There are several types of private line services offered by both common carriers and SCCs. These lines are separate from the public network and dedicated to the exclusive use of a particular subscriber. They should not be confused with the "private" numbers some executives have in their offices so that they do not have to use the regular company lines. Three private line services you should know about are tie lines, Off-Premises Extensions (OPX), and Remote Call Forwarding (RCF).

Tie Lines. A tie line is exactly what its name implies—a private line that ties two offices together across a street, across a city, or across a very long distance, and it allows unlimited calls between two locations for a flat fee. Tie lines are often dubbed *hot lines*. They are expensive, however, and can be justified only with a large amount of traffic between the two locations.

OPX. An off-premises extension merely ties an office in a separate building across a street or across the city into the company's main telephone system.

RCF. Remote call forwarding is not a private line in the truest sense of the word; rather, it allows a company to maintain a local number in a distant city so that customers do not have to incur long-distance expenses. For example, a company in Chicago may have many customers in Duluth. When the Duluth customer calls that local number, the call is automatically forwarded to Chicago on regular long-distance lines at no cost to the customer. The long-distance charge is paid by the company in Chicago subscribing to the RCF service.

□ Switched-Voice Services

Certain Specialized Common Carriers (SSCs) offer reduced-rate, long-distance services that compete with regular long-distance services offered by the common carriers who use AT&T's Long Lines. These companies have sprung up as a result of recent deregulation decisions by the FCC, and their services are known in the industry as *switched voice services*. Some familiar long-distance services are those offered by MCI and ITT, as well as "Sprint" and Western Union's "Metrophone." Switched-voice services are advertised for both consumer and business usage and offer savings only to those with high volumes of long-distance calls. If your company uses such services, you will recognize the following characteristics.[2]

1 You must use an access code to get into the system. For some services, a total of 22 digits must be dialed to complete a call, making automatic dialing devices very attractive.
2 Tone dialing is required. Adapters can be acquired for telephones with dials.
3 Most switched-voice services still reach only major metropolitan areas.
4 Some switched-voice services have minimum usage requirements.
5 All switched-voice services can reduce long-distance costs *only* if properly used.

■ RECORD COMMUNICATION SERVICES

Record communication is the term applied to a form of electronic mail in which messages are keyed into a carrier's network, transmitted by telecommunication paths, and delivered by telephone or in document form or both. The services are called record communications because of the record provided by the document. The most familiar record carrier is Western Union, and the most familiar record communication is the telegram, sometimes called a wire. The most familiar method for sending record communications is for the customer to call the Western Union office and dictate the message over the phone. A written or typed message can also be delivered to the carrier's office. The message is then transmitted by wire to the carrier's office nearest the addressee and delivered by telephone, messenger, or local mail, sometimes within a matter of hours. The message may even be picked up by the addressee. Charges for messages telephoned or delivered to the Western Union office appear on the sender's telephone bill.

[2] Arredondo, *Telecommunications Management*, pp. 115-117.

In the past, record carriers were divided into two categories, domestic and international. International Record Carriers are commonly known as IRCs. However, deregulation and technological developments in the communications industry are affecting record-carrier services just as they are affecting telephone services. The networks of some carriers now permit delivery both within the United States and to foreign countries.

☐ Message Telegrams

Domestic telegrams may be for same-day delivery or overnight delivery, which is less expensive. International telegrams may be either Full-Rate Telegrams (FR), which usually have same-day delivery, depending upon the destination, or Letter Telegrams (LT), with overnight delivery. International letter telegrams are often called *night letters*, and they are less expensive than full-rate telegrams. Overseas telegrams are sometimes referred to as *cables*.

☐ Mailgrams

Mailgrams are a special service offered by Western Union in cooperation with the U.S. Postal Service. Messages are transmitted by wire to the post office nearest the addressee and then delivered by regular mail. Next-day delivery is guaranteed with cost savings over regular telegraph service.

☐ Telex and TWX

Telex is an acronym for *Tele*printer *Ex*change. TWX is an abbreviation for *T*eletypew*w*riter E*x*change. Telex and TWX are similar services, but they operate on separate networks. These services differ from other record-carrier services because customers are subscribers; that is, they have teleprinters installed on their premises, which gives them direct access to any subscriber on the network. Subscribers are assigned calling numbers similar to phone numbers, and they can communicate through their teleprinters with each other any time, just as telephone customers do. Telex interfaces with worldwide teleprinter exchanges, while TWX is confined to the continental United States, with some subscribers in Canada and Mexico. Directories of subscribers are available, and subscribers receive monthly billings.

Because charges for record communications are based on the length of the transmission in minutes or the length of the message in words, messages must be carefully written for brevity. Some companies use special codes so that certain words may stand for whole phrases or even whole sentences. KALOP, for example, may stand for the following: We authorize you to act for us and will confirm by mail. Several standard codes are available, or a company may even devise its own private code both for security purposes and to reduce costs.

Record communications offer several advantages over voice communications for users with high volumes of short, routine messages, such as orders or price changes. These communications are faster than regular mail, but less expensive than telephone calls. In addition, a telegram or mailgram generally has greater psychological impact than a letter.

☐ Special Carrier Services

Although the Western Union telegram is the most familiar record communication, it is by no means the most frequently used form of record communication in business. TWX, Telex, and other record carriers, such as Western Union International (not connected with Western Union), ITT World Communications, Inc., RCA Global Communications, TRT Telecommunications, and FTCC Communications (French Cable), carry vast quantities of communications for businesses daily. All these carriers offer many or all of the following special services.

- *Delivery confirmation.* Deliveries can be confirmed automatically by the carrier's system so that the sender knows the message has reached the correct destination. Nondeliveries are also reported.
- *Store and forward.* Messages can be automatically stored in the carrier's computer system and then forwarded later when the addressee is available or the addressee's terminal is not in use.
- *Multiple-address messages.* The carrier's computer allows the user to key in a message one time and then instruct the system to send it to the list of addressees that follows. Similarly, a list of addressees can be stored for transmission of messages, such as monthly updating of price lists or inventory levels, at regular intervals.
- *Unattended operation.* Some teleprinters have features such as automatic redial and automatic shutdown that enable them to transmit without an operator present.
- *Compatibility with other systems.* More and more record-carrier equipment and services offer compatibility with computers, word processors, TWX, Telex, radio, and marine communications. This compatibility is vital in today's integrated office operations.
- *Special billing services.* The systems of some record carriers report the costs

of the transmissions immediately upon completion, when the operator inserts a special code. Others provide special itemization on billings for customers who want to allocate their costs among departments.

■ TELECONFERENCING SERVICES

Probably the telecommunications services with the greatest potential impact on business activities are the teleconferencing services. Not only do they reduce the cost of meetings by making travel unnecessary, but they also hold promise for increasing the productivity and decision-making capacities of management personnel by improving communications among people and timeliness of information flow. Teleconferencing activities fall into roughly three categories.

□ Computer Conferencing

Computer conferencing involves people using communicating computers to conduct meetings. Such meetings are prearranged and usually follow planned agendas. The participants do not necessarily talk with one another. Instead, presentations are made and participants exchange ideas from remote locations by way of computer input, which is transmitted over telephone lines and displayed on the CRT screens of all participants. A large data base can be accessed by the participants during the conference. For companies who do not have communicating computers, conferencing services are offered by firms such as Graphnet, Tymnet, and GTE's Telenet.

□ Graphics Teleconferencing

Graphics teleconferencing involves meetings by telephone, augmented with facsimile or visual displays of graphs, charts, or pictures as electronic images on television-like screens. The people cannot see each other, but they can see what they are talking about. The graphics display is often in full color and may be generated by a computer that can make vast quantities of information instantly available to participants.

□ Video Teleconferencing

Videoconferencing may be by slow-scan television (the picture changes only once each second) over regular telephone lines or full-motion television (the picture changes 33 times each second, so the motion appears normal) over more expensive wideband transmission paths. All participants can see each other, and graphics displays may be used.

One of the first firms to offer videoconferencing service was Satellite Business Systems (SBS). AT&T's Picturephone Meeting Service (PMS) is another example of videoconferencing. Meeting rooms are equipped with cameras, screens, and audio facilities. In addition, hardcopy machines produce copies of images displayed, and videotape recorders are available to record the meetings. Typical charges for the use of rooms when the service was inaugurated in 1982 were $1340 per hour from New York to Washington, D.C., and $2380 between New York and Los Angeles.

■ DIRECTORY SERVICES

The first telephone directory appeared with the first telephone exchange on February 21, 1878. It was a single sheet of paper, 6 inches by 9 inches, listing the names of the 50 subscribers to the telephone service. There were no telephone numbers then because dialing had not been invented, and addresses were not included. The directory also stated the

Developmental prototype facility of AT&T's Picturephone meeting service (PMS). Photo reproduced with permission of AT&T.

hours during which the exchange was open and calls could be made. Even then, however, some classifications were provided for business subscribers, with separate headings for physicians, stores and factories, meat and fish markets, and hack and boarding stables. By 1890, the number of telephone subscribers had grown to such an extent that a bound book was necessary, and the business section was already distinguished by its yellow paper.

Today's telephone books are essentially the same—even the huge, multivolume directories published for large metropolitan areas. The white pages are still an alphabetical listing of subscribers with the addition of addresses and telephone numbers. The Yellow Pages are business advertisements classified by product or service. However, some vital information has been added that makes telephone directories remarkably expansive resources for office workers, and most directories are available without charge.

Directories published by telephone companies may include information such as the following:

1. Special listings of emergency and community service numbers for quick reference.
2. First aid, disaster, and emergency instructions.
3. Complete descriptions of telephone company services and policies.
4. Consumer information and tips on how to save money and get the most from your telephone service.
5. Rate information for local and long-distance calls.
6. Area code, time zone, and zip code maps and prefix locations.
7. Perpetual calendars.
8. Special services for the handicapped.

If you need directory information for other locations, many telephone business offices maintain directories of other cities for your use. Most public libraries also maintain current directories for major cities. The reference librarian will check a listing and give you an address and phone number over the telephone. Libraries also maintain directories that are published by trade associations and private directory firms. These directories provide address information about companies, as well as names and titles of key personnel. Examples of such directories are the following:

- Thomas' Register of American Manufacturers
- Standard & Poor's Register of Corporations
- Jane's Major Companies of Europe
- Japan Company Directory
- Directory of American Firms Operating in Foreign Countries

Directory-assistance operators are able to look up numbers much faster than we are. Some have computerized directory systems, enabling them to call information up on a screen almost instantly. Everyone tends to take excessive advantage of this service in spite of the polite recordings we hear. Some telephone companies are charging for directory assistance calls after a certain allowed limit has been exceeded each month. You should be careful not to add to your company's phone bill by calling the operator when a directory is at your fingertips.

■ CONSULTING SERVICES

The telephone equipment and services in your company are carefully planned to do the best possible job at optimum cost. The primary concerns in the design of the plan are the overall requirements of the entire operation, not just your individual needs or the needs of your department. As you can surmise from what you have learned thus far, there are a lot of factors to consider.

Consulting services are available to help the executives of your company make good planning decisions regarding the entire communications system. Most telephone companies offer the services of customer service representatives for this purpose. These customer representatives are frequently backed up by a team of experts with specialized knowledge of both telecommunications and the specific needs of a particular industry.

If your company purchased its equipment directly from a manufacturer, one important consideration in that purchasing decision was the quality of consulting and maintenance service offered by the vendor. In addition, your company may employ a telecommunications specialist or retain one of the many private management consulting firms who specialize in telecommunications.

Whatever type of consulting service is used to maximize the effectiveness of your telephone service, it is your responsibility to cooperate fully with the consultant or staff member. Part of your job is to understand the operation and maintenance of the equipment at your work station and to use it properly, to provide accurate information about your use of and experience with your equipment, and to be receptive to the changes that may be implemented in the system.

■ MAINTENANCE SERVICES

If something goes wrong with your equipment, you should consult your supervisor immediately. It is most likely that the telephone company will be responsible for repairing it. In some cases, however, the vendor may be called, or your company may have maintenance people on staff.

■ TRAINING SERVICES

Carriers and equipment vendors in the telecommunications industry offer a wide variety of educational training services and publications for customers, business users, and classroom students. In addition, films and publications are also offered by professional associations serving the industry. Some of these associations are listed in the back of this book.

As mentioned in the chapter on equipment, vendors provide training for console operators and users at the time equipment is purchased and sometimes on a continued basis. Detailed training and user manuals are also provided by vendors. These services are generally free of charge.

Some telephone companies have simulated telephone models of a six-button key system, called *business telephone trainers*, which are available for loan or demonstration. These allow students or office trainees to experience incoming calls and to transfer calls under the direction of a supervisor or trainer without using outside lines. There are also simulations on cassette tapes, which allow participants to practice talking on the telephone by interacting with a recorded program. They can then play back the recording and analyze their own performances.

Films, filmstrips, tapes, booklets, and pamphlets are available on such subjects as telephone manners, tips for economical and effective telephone usage, the role of the telephone in community and emergency services, telephone technology, and telephone industry careers. Most are offered without charge.

Finally, telecommunications is now a management specialization in its own right, and there are training opportunities for those aspiring to careers as telecommunications managers, consultants, or specialists. Courses, seminars, and conventions are sponsored by companies, as well as by trade and professional associations. In addition, several colleges and universities in the United States offer certificate, diploma, and degree courses in telephone technology and telecommunications management.

Any time your company provides training programs or retains consulting services such as these, the purpose will always be the same: to effectively combine telephone equipment with telephone services into a system that enables people to meet the objectives of the business. We discuss some of these systems, large and small, in the next chapter.

QUESTIONS FOR REINFORCEMENT AND DISCUSSION

1. How can long-distance calls be useful to a business?
2. Tell how the following long-distance services can save money.
   a. DDD
   b. WATS
   c. Rate periods
   d. Switched-voice services
3. Describe three long-distance operator services and tell when they might be used.
4. How can employees make personal calls on the office phone so that the company does not pay for them?
5. Between what times can you place a long-distance call and have both parties talking when the windows are open in both cities?
   a. From San Francisco to New York
   b. From Miami to Denver
   c. From Houston to Fairbanks
   d. From Atlanta to Salt Lake City
   e. From New York to Tokyo
   f. From San Diego to Paris
6. How can the public library help the telephone user?
7. What should you do if you make an error dialing a long-distance call?
8. What should you do if you get a bad connection for a long-distance call?
9. What are some important misconceptions about WATS service?
10. What are some important characteristics of switched-voice services?

TRUE-FALSE QUESTIONS

Write T *if the statement is true and* F *if it is false.*

_______ 1. A foreign exchange means an exchange located in another country.

_______ 2. A tie line connects two locations on a private telephone line.

_______ 3. You do not have to pay to find out a telephone number in another area code.

_______ 4. WATS is free.

_______ 5. Operator-assisted calls are always more expensive.

_______ 6. It does not matter which line you use as long as you are calling the same area code.

_______ 7. WATS is always cheaper.

_______ 8. Only telephone companies publish directories.

_______ 9. Some companies would not save by subscribing to switched-voice services.

_______ 10. A collect call is never accepted in a business office.

USING THE TELEPHONE DIRECTORIES

Using your local telephone directory, either find or tell where you would find the following information.

1. The area code for Okmulgee, Oklahoma.
2. The meaning of FX.
3. The correct time.
4. The current weather report.
5. The nearest office supplies vendor.
6. The phone number of the nearest office of the Federal Civil Service Commission.
7. The zip code for your branch office.
8. Long-distance rates between your city and London.
9. The phone number of the Rescue Squad.
10. The meaning of an abbreviation on your phone bill.

RECORD COMMUNICATIONS

Choose the correct answer for Questions 1–9 and write your answer for Question 10 on a separate sheet of paper.

1. A telegram is a form of record communications because:
   a. It is used primarily by record companies.
   b. It is transmitted by wire with a written document provided for the receiver.
   c. There is always a record of the transaction.
2. An important advantage of record communications is:
   a. Messages can be delivered faster than by regular mail.
   b. Messages have a greater psychological impact.
   c. Both of the above.

3. An important disadvantage of record communications is:
   a. A telegram is so small that it might get lost.
   b. Record communications are not economical for long messages.
   c. Only companies with special equipment can send record communications.
4. Companies that subscribe to Telex or TWX services can:
   a. Send messages to anyone in the world without Western Union.
   b. Compete against Western Union in the record communications business.
   c. Send messages directly to other subscribers on the network.
5. Companies who provide international record communications services are known as:
   a. CRTs
   b. IRCs
   c. FXs
6. Charges for wired communications are based on:
   a. The distance the message travels.
   b. The time of day the message is sent.
   c. The length of the message in words or the transmission in minutes.
7. Mailgram is a service offered as a joint effort of:
   a. Western Union and and the phone company.
   b. Western Union and the United States Postal Service.
   c. Western Union and Eastern Communications, Inc.
8. Teleprinters used in connection with Telex and TWX can be compatible with:
   a. The company's data processing equipment.
   b. The company's word processing equipment.
   c. Both of the above.
9. *Cable* is another term for:
   a. Special wires used only for record communications.
   b. An international telegram.
   c. Computer Assisted Basic Letter Entries.
10. Rewrite the following message so it will be a more economical record communication.

    The items ordered on your Purchase Order No. 4962 dated June 16, 1983, will be ready for shipment on August 29. Please let us know the method of shipment you prefer and which of your New York offices should receive them.

5

Systems

In this chapter, you will take your first step into the office of the future by changing your title. Instead of an office worker, you are now called a knowledge worker, because this title more aptly describes the work you do. You are a very important part of the Information Revolution discussed in Chapter 2. Another very important part of the Information Revolution is your telephone. You and your telephone function in an integrated office system that captures, processes, distributes, stores, and retrieves the knowledge required by every person in your company. This is true whether the company is large or small, and it is true whether the knowledge you work with is a routine telephone message or the data concerning fuel to be used for launching a spaceship.

Studies have shown that managers and knowledge workers spend more time involved in communication than in any other activities. Communication activities include face-to-face discussions and conferences, written communications, and telecommunications. The amount of communication conducted over telephone paths will vary from business to business and depend on the size of the operation and the nature of its activities.

The larger the operation and the more complex its needs, the more likely that a telecommunications manager will be hired. The responsibilities of such a person are as follows:

- To understand all phases of the business operation, as well as the industry in which it functions.
- To have a thorough knowledge of telecommunications equipment, services, vendors, and carriers.
- To evaluate the company's system and plan effectively for changing needs.
- To maintain the system effectively and control the costs of telephone usage.
- To keep abreast of technological developments that may offer improved equipment and services.[1]

The job of the telecommunications manager is more complicated today than it has ever been before. As you will see, the 100 million telephones visible on the desks of American offices are but the tip of the iceberg that is the vast communications network feeding the information appetites of people, business, and government.

In order for you to learn about some of the systems and applications of telecommunications in business, you must first understand more about the different kinds of telephone paths and about the four types of telephone communications, which are voice communications, data communications, facsimile communications, and video communications.

TELEPHONE PATHS

The most common telephone paths are narrowbands and voicebands, the wires and cables you see strung from pole to pole and the wires and cables you do not see underground and under water. Narrowbands are the slowest conductors and are used for some TWX and private line teletypewriter transmission. Voicebands are used for most of the voice communications throughout the world and for a great deal of data, facsimile, and slow-scan video communications.

When wires are combined into cables, the results are paths with larger and faster transmission capacities. One of the most famous cables is the coaxial cable, which consists of up to 20 copper tubes, each about as big around as a pencil. A pair of these tubes can carry up to

[1] Larry A. Arredondo, *Telecommunications Management for Business and Government*, 2nd. ed. (New York: The Telecom Library, Inc., 1980), pp. 12–14.

3600 telephone conversations. More recent technological advances permit greater speeds and capacities with much smaller wire and cable, which requires less space, is less costly, and is easier to install. These paths are sometimes called widebands, or broadbands. One of the most-important developments in cable is fiber optics. Instead of wires inside the cable, there are very fine glass or plastic threads; the signal travels over these threads as a beam of light, which may be powered by a laser. Transmission can be as much as 1000 times faster and much more accurate with fiber optics than with wire.

Some telephone transmission paths use no wire or cable at all. They are microwave radio beams. Microwave transmission has a disadvantage because the distance a signal can travel is limited. Therefore, repeater stations must be placed about every 30 miles on top of buildings, towers, or mountains. In satellite communications networks, the repeater station is placed on a satellite, usually orbiting the earth geo-

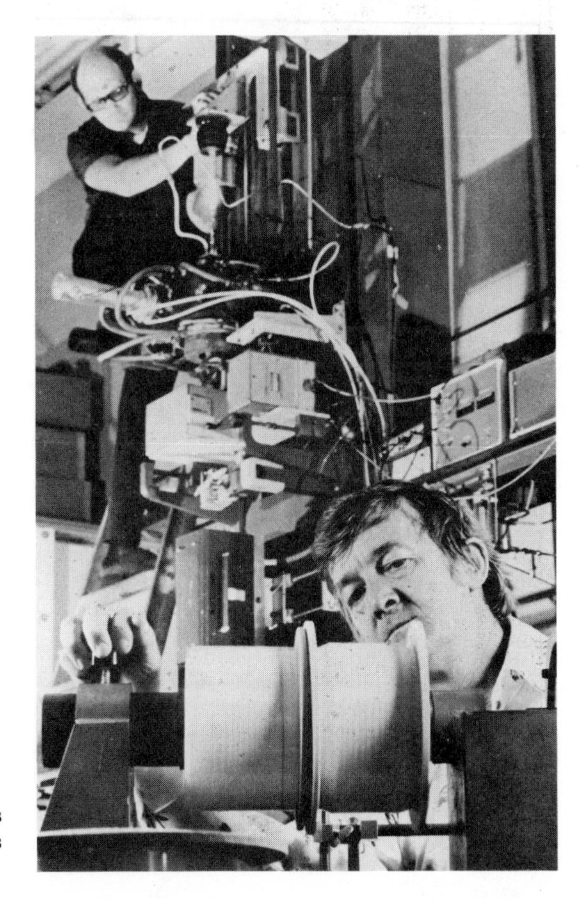

(a) Manufacture of glass fibers for light wave communications systems.

(b) A light wave communications system, by which messages are communicated by light pulses.

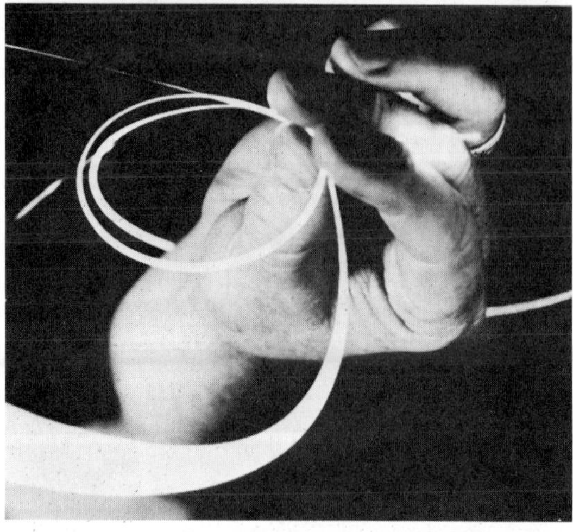

(c) Ultrapure glass fibers serve as "wires" for future telecommunications transmission.
Photos reproduced with permission of AT&T.

synchronously, which means it stays above the same location all the time. A signal can actually reach its destination faster by way of a satellite 23,000 miles in space than it can traveling across land paths.

☐ Voice Communication

Voice communication is the fancy term used for people talking on the telephone. Voice communication systems can be simple POTS (Plain Old Telephone Service, with two people talking on single-line telephones) or elaborate combinations of the various equipment and services described in the previous chapters. The great majority of business and government offices use only voice communication services.

☐ Data Communications

Data communication is the term used for machines talking on the telephone. Information is converted into electronic signals and sent over telephone paths. Data communications usually involves computer equipment used to process and send extremely large volumes of information at extremely high speeds over very long distances. Both numbers and text may be transmitted by data communication systems. These systems are discussed again later in this chapter.

☐ Facsimile Communication

Facsimile communication is the transmission of images over telephone paths, usually by telecopiers or communicating photocopiers such as those discussed in the chapter on equipment. Software for creating compatibility between facsimile equipment and other business machines such as Telex, TWX, and word processors is now available, which enables companies to integrate FAX into their communications system.

☐ Video Communication

As its name implies, video communication involves the transmission of television signals over telephone paths. Such equipment allows people to talk and see each other at the same time. Videoconferencing services,

3M's new EMT 9140 digital facsimile transceiver permits an operator without special training to send or receive exact copies of business and other documents to or from anywhere in the world in seconds via standard telephone lines. Photo courtesy of 3M.

such as those offered by AT&T and SBS, are relatively new on the business scene and are rapidly gaining acceptance.

■ SYSTEMS FOR SMALL BUSINESSES

In a small office, such as an insurance agency serving a local community, the telecommunications needs may be adequately served by a six-button key system consisting of one trunk and two or three rotaries, plus an intercom path. The intercom may serve six or eight stations, all of which have identical telephone sets. A secretarial knowledge worker may answer the phone and route calls from any station, right along with other secretarial duties. Toll and long-distance calls may be important in the company's activities, but there are not enough to justify any of the special services, such as FX, WATS, or switched-voice services. Such a system may be considered to be the simplest of office telephone systems.

Even small firms, however, may have a great deal of long-distance traffic. An independent wholesale merchant may deal with a group of manufacturers and retail outlets spread over a geographic area encompassing several states. Such a merchant may have one office and a small staff, but require a private line to the main supplier and several FX lines

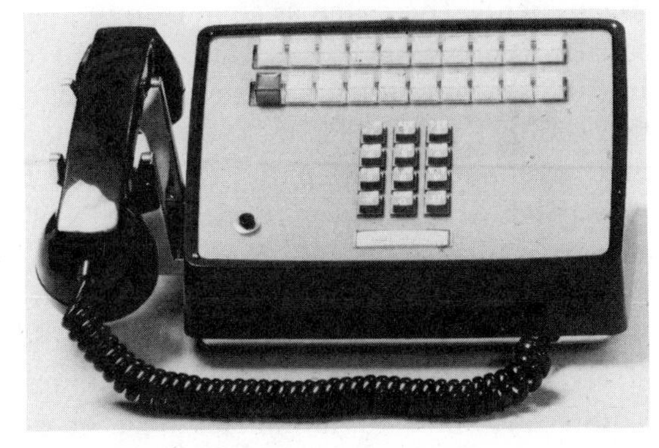

Many of today's small offices are well served by the workhorse of business systems, the key telephone, shown here. Photo reproduced with permission of AT&T.

to areas where customers are located. If the merchant is a nationwide distributor, several WATS lines may be in order—IN-WATS for customers to call toll-free and OUT-WATS for salespeople to generate business. On the other hand, if the merchant has many short messages, such as routine orders, TWX or Telex service may be even more economical than voice communications. At this point, the telecommunications system of even a small operation becomes somewhat complex and requires a thorough understanding on the part of all the users in the office.

■ SYSTEMS FOR MEDIUM-SIZED COMPANIES OR OFFICES

Medium-sized operations have needs that are much more varied than smaller operations. Consider, for example, an office supply house serving a major city. The system of this firm must serve a store, a large warehouse, a telephone order department, an office staff, and perhaps a staff of equipment-repair personnel, who are nearly always in the field. An electronic key system with several trunks serving 30 or 40 stations may now be required. The stations serving the telephone order department may be programmed for ACD, which—you will remember—is the automatic distribution of calls coming in from customers among order-taking personnel. Reprogramming of the ACD system may be necessary during busy seasons or advertised sales so that stations can be added to the group to handle the increased volume of calls. The system's paging feature may be tied into a network of loud speakers in the warehouse and on the retail sales floor, and remote paging devices may be carried by repair personnel in the field.

Another application of an electronic key system may be in a government office such as a courthouse, where attorneys and judges are constantly moving between offices and courtrooms. Call forwarding and message-recording services may be vital to efficiency. Administrative support personnel may have so much interoffice communication that several intercom paths are needed. Duties may be assigned and reassigned with varying caseloads so that frequent reprogramming is necessary to keep up with changing needs.

A larger system may be required by a regional textbook-distribution center serving a large number of widely dispersed schools in a territory of six or eight states. Efficient communications for this office may require FX lines to minimize the cost of local calls within its large metropolitan location, tie lines to headquarters and other regional offices, IN-WATS lines for calls from customers, and switched-voice services for outgoing calls made by executives and sales personnel. This office may have several hundred employees who use—and potentially misuse and abuse—these very costly long-distance services. In fact, it would hardly be reasonable to expect all these employees to have sufficient knowledge of the system to choose the least-expensive path for every call, and it would not be entirely logical to assume the firm would not be a target for some abuse. This company could probably reap considerable benefit and control by installing its own intelligent switch on the premises.

Such a system would provide LCR, which—as you remember—automatically chooses the least-costly path as the call is being dialed. The electronic switching system could also be programmed to restrict the class of service for certain stations so that no long-distance calls could be made from those phones. The Call-Detail Recording (CDR) feature on the switch could also report exactly what calls are made from each station, so that costs could be allocated properly and, if necessary, abuse could be detected and corrected.

Special demands are placed on the telephone systems of hotels and motels. An independent study by Panell, Kerr, and Forster published in *Trends in the Hotel Business* showed that hotels can lose as much as $140 per room per year due to guest usage. Most hotels restrict the class of service for stations in guest rooms so that long-distance calls are routed through a toll operator and charges are added to the customer's bill. However, recent changes in legislation make it legal for hotels and motels to earn a profit from guest telephone usage. Therefore, LCR and CDR capabilities are becoming more attractive. Telephone service can then be provided to guests with maximum convenience at the lowest possible cost to the hotel. Charges can be printed on guest bills immediately, and accounting reports are produced monthly.

An important trend for very large hotels in major metropolitan areas is the installation of satellite-delivered, videoconferencing facilities for meetings and conventions. Meeting rooms are equipped with graphics display terminals and large television screens linking them with other hotels in the chain, convention centers, corporate offices, and television studios.

Hospitals are like hotels in that telephone service is often provided for patients. More important, however, is the fact that hospitals require special telephone links for emergency services. The paging systems and mobile units, usually operator-accessed, provide instant communication channels that help medical personnel save lives. In addition, hospitals use FX lines to other hospitals and clinics, OPX lines to offices of staff and doctors, and tie lines to specialized treatment facilities and disaster units.

Finally, teleconferencing enables medical specialists to communicate with one another from remote locations; and data communication systems such as those described in the next section of this chapter are enabling medical centers, hospitals, clinics, and private doctors to share a vast data base of medical information recognized to be a major contribution to successful practice of medicine and progress of medical science.

■ SYSTEMS FOR LARGE COMPANIES AND OPERATIONS

The remainder of this chapter is devoted to systems that serve the needs primarily of very large operations and deserve more detailed description. They are data communications, electronic mail, integrated office systems, and network control.

□ Data Communications Systems

When computers talk on the telephone, the equipment used is considerably more complicated and the process is considerably less casual than when people talk on the telephone. In fact, personnel involved in data communications systems must possess a great deal of highly technical information that knowledge workers do not really need. Because data communication is often a very important part of a large office operation, however, you should understand some of the concepts. A basic model of a data communications system and a discussion of its components is shown in the figure below.

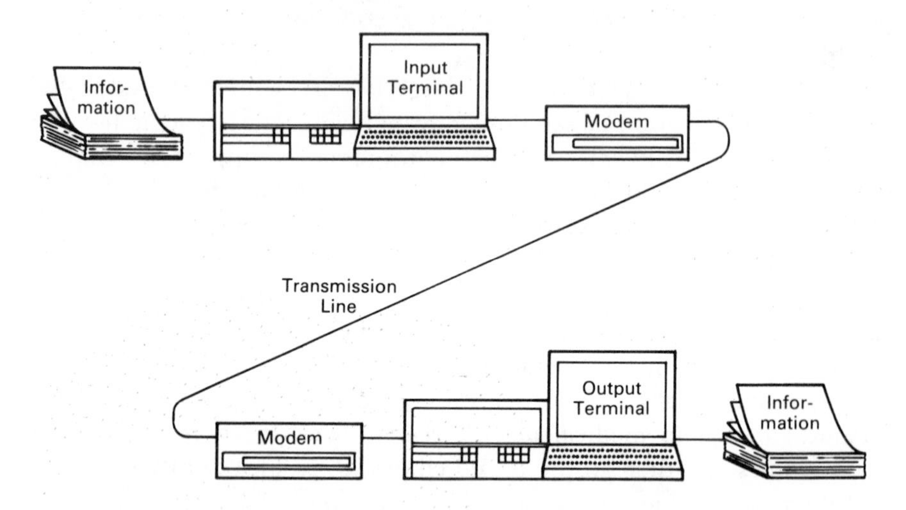

Input and Output Devices. The message must first be fed into the system at the sender's location. This is usually done by typing or keystroking the information on a teletypewriter or at a computer terminal. The input terminal encodes the message into symbols for transmission. After transmission and decoding, the message is displayed at the receiver's location. This may be accomplished on a CRT screen or by a teleprinter, which produces hard copy.

Binary codes are used in data communications. *Binary* means that only two symbols are used. The two symbols used are the digits 0 and 1, and they are called *bits*, which is an abbreviated combination of the words *binary* and *digit*. Every letter, number, character, and instruction in the code is made up of some combination of 0 and 1. One commonly used standard code is theAmerican Standard Code for Information Interchange (ASCII), in which each character, letter, number, or instruction is made up of seven bits. In the ASCII (pronounced "askie") code, the letter *A* is 1000001. The instructions in the bit stream are called *protocols* and they give the equipment instructions, such as where the message is going, how it will get there, where it begins and ends, and when it is to be checked for errors. There are a number of protocols in use, such as the IBM Synchronous Data Link Control (SDLC).

Modems. Once the message is encoded and fed into the system, it must be converted into a signal that can be transmitted over telephone paths. In telecommunications, this conversion process is called *modulating*. At the receiver's location, the reverse process is called *demodulating*. Therefore, the name of the device that performs these functions is an abbreviated combination of these two words: *mo*dulate + *dem*odulate = *modem*.

Transmission speeds are stated as the number of bits sent per second (bps). Ordinary telephone lines (voicebands) can handle about 4800 bps. Widebands can handle bit streams in the millions of bits per second. At such high transmission speeds, the entire contents of a 24-volume encyclopedia can be transmitted in 3 minutes!

Transmission. Both common carriers and SCCs offer data transmission services. Common carriers include the Bell system, independent telephone companies, and Western Union. The emerging SCCs are offering facilities and services more suited to the requirements of data communications. Networks of some SCCs include satellites and minicomputer nodes, which offer highly sophisticated services and capabilities; such carriers are known as Value-Added Carriers (VACs). Their networks are called Value-Added Networks (VANs).

The transmission paths may be either a dedicated or a dial-up service. A dedicated serivce is a leased private line, while a dial-up service is used by many subscribers who dial into the network as though they were making ordinary telephone calls.

Error Detection. Data communications can be affected by poor telephone connections, just like voice communications. When this occurs, data received may contain errors. Error detection may be accomplished by one of the following three methods:

1. A human may catch the error and request retransmission.
2. The data may be transmitted back to the sending location for verification, a process known as an *echo check*.
3. A group of bits on which a mathematical calculation is performed may be transmitted. If the answer to the calculation is the same at the receiving location as at the sending location, the data is assumed to be correct. This is often known as a *parity check.*

Security. To protect the privacy of data transmission, some systems scramble the message during transmission or use a device to convert the message into a code that is not standard. This secret coding is known as *encryption*, and the device is an *encryptor*.

Two familiar applications of data communications are in airline or hotel reservations and in automatic bank tellers. Other applications are for (1) customer account information, perhaps for a national oil company or for a bank credit card service, (2) inventory control systems for firms with widely spread warehouses, (3) libraries with many branches in a large city, or (4) medical records systems for a group of hospitals and clinics. With data communications systems, these users can have almost instant access to information even from very remote locations.

□ Electronic Mail

Computers are not the only office machines that talk on the telephone. Word processing machines and photocopiers can also communicate over telephone paths. In fact, since word processing and data processing began to merge, an entire concept—electronic mail—is developing. There are four forms of electronic mail. You already know about two of them: facsimile equipment, which is discussed in Chapter 3, and record carrier services, discussed in Chapter 5. The two types of electronic mail discussed here are communicating word processors and Computer Based Message Systems (CBMS).

Communicating Word Processors. Communicating word processors most closely represent the automation of conventional mail. Letters are sent without the postal service. Here is how it works.

The Word Processing (WP) operator keys the message into the memory of the WP machine. If the originator wants to review the message before it is sent, a hard copy draft can be printed or the document can be displayed on a screen. Revisions or corrections are keyed into the machine and the memory can be corrected automatically. When both the originator and operator are satisfied with the message, it is transmitted to a word processor at the receiving location over a network of modems and telephone paths similar to those described for data communications. There is no need for postage or envelopes, and there is no waiting for postal service pickup and delivery.

The WP machine at the receiving end prints out the message for the person to whom it is addressed. Multiple copies can be printed for messages directed to more than one person. The message can be stored on an external medium (usually a magnetic disk) and reused without

Siemens T4200/40 text terminal combines the essential features of a word processor, an electric typewriter, and a teleprinter. This allows it to interface with communicating word processors, as well as with record carriers and data communications services. Photo courtesy of Siemens Corporation.

the necessity for rekeying. This procedure, known as *boilerplating*, allows the company to store frequently used form letters and send them as required with a simple instruction to the system. Some form letters, such as monthly sales reports, are updated and sent at regular intervals.

Computer Based Message Systems. CBMS takes the concept of electronic mail one more step up the ladder of automation by eliminating paper almost entirely. It is sometimes called the paperless office, and its users send and receive messages from electronic work stations, which are essentially computer terminals with screens. The message is keyed into the system by the sender and transmitted by telephone paths to the receiver's station, where it appears on the screen. Two advantages to users are: (1) messages are stored in the computer until the user is ready to receive them, and (2) messages become part of the data stored in the computer. This means that the sender can refer to any information in the data base and the receiver can retrieve and review it without retyping and without handling paper. Such systems are particularly valuable to engineers and scientists, who work with large amounts of technical data and information. Portable terminals, which allow users to access the system from ordinary telephones anywhere, are available.

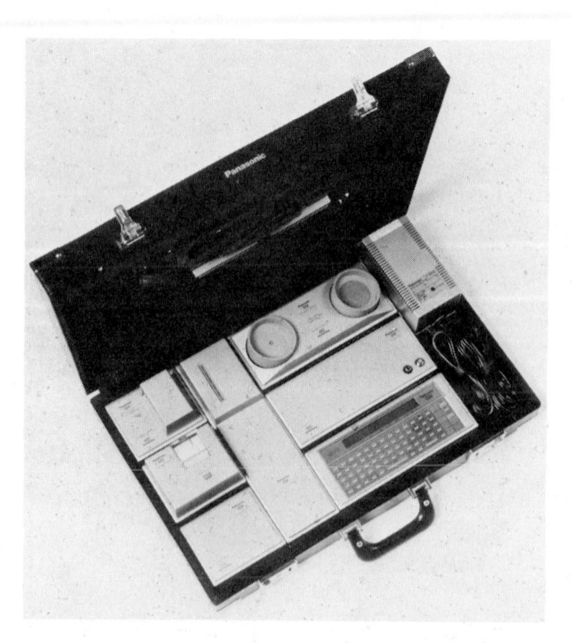

The Link,(TM) pictured here with its components, is a portable, intelligent computer terminal that enables a user to access a computer by telephone from any remote location. Photo courtesy of Panasonic.

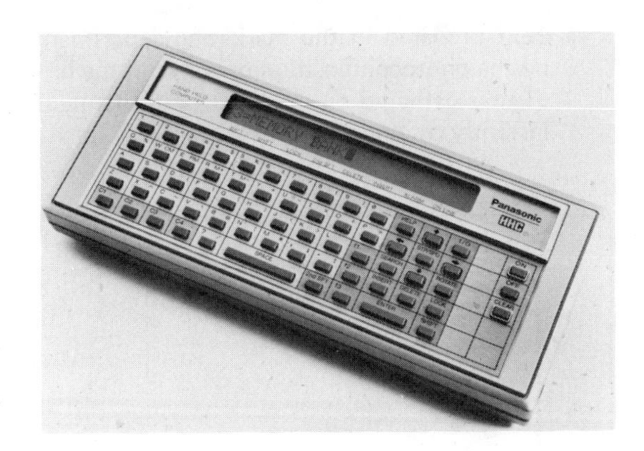

This portable computer terminal is the principal component in the Link™. Photo courtesy of Panasonic.

The primary advantage offered by electronic mail is the speed with which data and text can be distributed over telephone paths. Equally important, however, is the reduction of the use of paper. When you consider that postage rates are increasing, the cost of paper is rising at a rate of 12 to 15 percent per year, and the cost of physically storing and handling paper is at an all-time high, it is easy to see how electronic mail can effectively reduce communication costs. In addition, when you consider that the majority of documents stored on paper in the file cabinets of American offices are never used again (one firm did a study which revealed that only 5 percent of the paper in its files was ever used again), it is easy to see how the paperless office can significantly increase office productivity.

□ The Integrated Electronic Office

Integrated office functions mean everything is tied together, with resulting improvements in effectiveness and productivity. Very often, the information product of one office, department, or branch is an integral part of the product or function of another. Consider the following examples:

- Customer account information in a regional office must be included in the accounting reports of the headquarters office.
- Information submitted by department heads can be included verbatim in the monthly report of the general manager.
- Scientific data produced by the research department becomes part of the engineering specifications of the manufacturing department.
- Test results in product development become part of advertising text in the marketing department.

- Text prepared in the marketing department becomes part of the copy used by the photocomposition operator in the printing department.
- Figures gathered on market conditions in European branch offices are required instantly by financial officers throughout the United States.

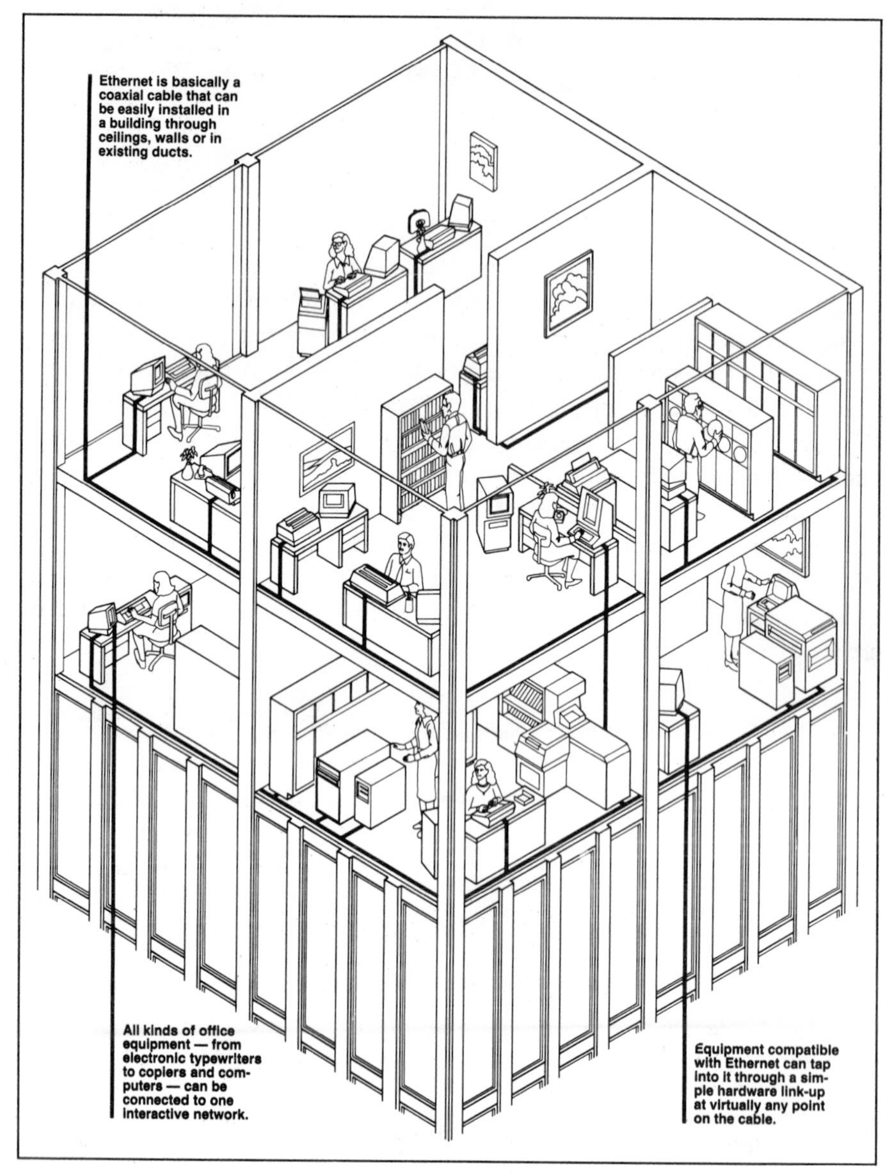

An integrated office tied together with the Ethernet cable. Illustration courtesy of Xerox Corporation.

As workflow progresses, all this information must be communicated again and again from office to office and from person to person. The more it is communicated by telephone, the faster the work is accomplished. If it must be recopied and rekeyed into word or data processing and record or data communications equipment, a great deal of duplication of effort results. On the other hand, if the equipment of one department communicates directly with the equipment of another office, data is inputted only once. Information is distributed among people quickly and expediently and stored automatically for future use. Time, work, and paper are minimized.

Most important of all, knowledge workers are freed from tedious, repetitive tasks for more challenging responsibility. Administrators and managers are able to make decisions and take actions based upon accurate and up-to-the-minute information, and human potential is maximized.

Systems for accomplishing this purpose are currently on the market. One such system uses Ethernet, a product of Xerox Corporation. Ethernet is a coaxial cable that is installed throughout the office building just like electrical wiring; it contains outlets, where machines such as computers, word processors, copiers, and printers can be plugged in. The Ethernets of several office locations can be interconnected by telephone lines. Unlike electrical wiring and outlets, however, this coaxial cable carries electronic signals from one machine to another, allowing them to communicate without human intervention. Thus the information product of one office or station can be directly accessed by another office or station without rekeying.

In order for such a network to be effectively utilized, the flow of work and communication needs of the knowledge workers must be carefully studied and thoroughly understood. It is much easier to install machine interconnections than it is to understand communication interconnections required by people. However, once the flow of work and communication needs of an organization are understood and systematized, an information carrier such as Ethernet can certainly enhance everyone's efforts.

□ Network Control

By now you probably realize that the telecommunications network of a large corporation may be quite vast and complex. In order for it to be administered properly, it must be continuously monitored, analyzed, and maintained. More and more corporations are paying very close attention to communications management and network control.

For purposes of control, the network may be divided into five parts: terminal equipment, intrafacility networks, local networks, long-haul networks, and international networks.

1 Terminal equipment includes every device that is connected to the transmission networks, from the ordinary telephones to the facsimile units, word processors, teleprinters, and computers.
2 The intrafacility network is the transmission network for connecting the offices and buildings of a firm. It may be simple intercom paths or it may be a network for the integrated electronic office, such as the Ethernet described earlier. Elaborate machine communication networks are often called Local Area Networks (LANs).
3 The local network connects the intraoffice network to the telephone company's central office or local exchange. A very large company with several buildings located in different areas may interconnect with several central offices and even several telephone companies.
4 Long-haul networks grow to be very complicated. They include every circuit or alternative path the user may have for long-distance voice communications (DDD, FX, OPX WATS, tie lines, and switched-voice services), as well as the paths dedicated to data, video, or facsimile communications. Some nationwide firms establish their own private networks by leasing lines from carriers for their exclusive use. Some even build their own transmission facilities, including microwave relay stations and satellites, and interconnect with a variety of carriers.
5 International networks include all four forms of communications, which may interconnect with networks of international carriers.

These components of a communications machine require constant planning, evaluation, and attention. Usage and maintenance of hundreds of pieces of equipment must be coordinated among vendors, and thousands of interconnections must be coordinated among carriers. Network controls must ensure that all circuits are transmitting properly, both for incoming and outgoing traffic, and that all circuits are properly utilized and not sitting idle. Diagnostic systems of modern switching equipment provide visibility so that managers can see where problems occur in the system and either provide maintenance or contact the responsibile vendor or carrier. Costs must be fairly allocated to users within the firm, according to good accounting practices. Finally, controls must be in place to ensure against misuse and abuse.

Moreover, people are constantly requesting new communications services, and organizational changes are constantly creating new demands on the system. Above all, the communications needs of the managers and knowledge workers must be recognized and satisfied throughout the ongoing process of communications management.

A good illustration of a corporate-wide communications system designed with all these objectives in mind is COMNET, which is the name given the system that serves over 100,000 employees of Rockwell International in its 300 facilities located on six continents. In 1982, COMNET included 80,000 installed telephones, over 1000 access lines, 300 InterMachine Trunk (IMTs) and 800 WATS lines, with plans for a total of 50 electronic switches.

Unique to the system is the nationwide backbone network of leased private lines controlled by four *d*igital *t*andem *s*witches (DTSs), also known as computer *nodes*, located in Pittsburgh, Pennsylvania; Cedar Rapids, Iowa; Dallas, Texas; and Seal Beach, California. This private, long-haul network is monitored by the Network Management Center (NMC) in Seal Beach. It carries most of the internal communications between Rockwell installations throughout the United States, and it interconnects with public long-haul facilities. In addition, it can interconnect with a variety of wideband transmission facilities for teleconferencing, data communications, and electronic mail. Rockwell executives spent 3 years studying their communications needs and available services, facilities, and equipment before making a decision and phasing in the system. This phasing in will continue throughout the 1980s and includes plans for building some privately owned transmission facilities.

Central to the control features is the system of authorization codes, which are assigned only to employees who are authorized to use the system; these codes must be inputted after the telephone number has been dialed. Regardless of which station the employee may be using, the computer recognizes the employee's identification number and class of service. In addition, each employee with an authorization code receives a monthly "phone bill," or printout of all calls made.

Rockwell officials report that the planning and installation of the system thus far have paid big dividends in cost controls, system maintenance, and user satisfaction.

QUESTIONS FOR REVIEW AND DISCUSSION

1. What is the product of the office?
2. What is the objective of a telecommunications system?
3. Give a specific application for each of the following types of communications.
   a. Voice communications
   b. Data communications
   c. Facsimile communications
   d. Video communications

4. Describe some of the special communications needs you think might be found in the following companies or industries and identify the system or service that would meet that need.
   a. Hotels
   b. Hospitals
   c. Convention centers
   d. Large franchise operations
   e. Financial institutions
   f. Multinational corporations
5. Describe the data communications model.
6. What is the meaning of the following terms?
   a. Encoding
   b. Decoding
   c. Modem
   d. Parity check
   e. Echo check
   f. Encryption
7. What specific data or information do you think might be transmitted for each of the following data communications systems?
   a. Airline reservations
   b. Inventory control systems
   c. Bank credit card service
   d. Automatic bank tellers
8. Cite an example of electronic mail and tell how the telephone makes it possible.
9. What is the meaning of each of the following commonly used acronyms?
   a. CRT
   b. WP
   c. DP
   d. LAN
   e. OCC
   f. SCC
   g. VAN
   h. EDD
   i. PBX
   j. CBMS
   k. ASCII
   l. POTS
10. Give a complete description of each of the following systems.
    a. Automatic call distribution (ACD)
    b. Least cost routing (LCR)
    c. Call-detail recording (CDR)
    d. Directory lookup systems (DLS)
11. Give some examples of how the information product or output of one office becomes the input of another office.
12. How can the integrated electronic office facilitate workflow where human communication interconnections occur?
13. Identify each of the following components of a communications network.
    a. Terminal equipment
    b. Intrafacility network
    c. Local network
    d. Long-haul network
    e. International network
14. What are some of the concerns of network controls?
15. What do you think will be the most important contribution of today's telecommunications technology?

6

Telephone Techniques

Using the telephone for a business purpose (or for any other purpose) should be just as easy and effective as face-to-face communication. Most of the techniques for successful telephone conversations are based on the same common-sense principles of human relations that promote good communication in any situation.

■ YOUR TELEPHONE PERSONALITY

The most readily perceived aspect of your telephone personality is your voice. Modern technology assures that your voice will sound the same over the phone as it does in person, so you can use your natural voice for telephone conversations. However, you cannot rely on nonverbal signals such as facial expressions or gestures. You must communicate your ideas entirely with what you say. It is a good idea, therefore, to speak at a slightly slower speed and to pay extra attention to pronouncing each word clearly and distinctly. Keep in mind that a slightly lower tone is usually more pleasing to any listener. The volume, or loudness, of your voice should be the same as the volume you would use for a person sitting across from you at a desk.

The most important aspect of your personality is the reflection of your attitude. Even without any visual contact, your attitudes, moods, and feelings will travel across the wires. Your smile can, in fact, be heard. To project the best attitude, you should assume both the correct physical posture and the correct mental posture.

Correct physical posture begins with the position of the handset. The mouthpiece should be about 2 inches (or the width of two fingers) from your mouth. Special telephone equipment, such as a headset or a Speakerphone®, is designed or installed so that the transmitter is automatically in the correct position. However, it is up to you to keep the transmitter in the handset within proper range, even if it is equipped with a shoulder rest. You should also sit up straight in your chair and avoid resting your chin on your fist or your forehead in the palm of your hand. If you permit your transmitter to stray up above your forehead or down under your chin or if you allow any object to block the transmitter, it is as though you have turned your back on a person conversing with you.

Correct mental posture means keeping your attention on the call by pushing distractions out of your mind and other tasks out of your way. You will improve your concentration if you designate a special area of your work station for your telephone and notepad. Then, shift all of your attention to the call. Do not shuffle through papers on your desk. Avoid carrying on a "conversation" with someone present by silent gestures or scribbled notes. Poor attention is the most frequent cause of poor telephone manners, and good attention can be achieved only with conscious effort, especially in a busy office.

You must, in fact, give the call the same total attention you would if the caller were present in your office. Try to get a genuine appreciation for the person who called you, and be sure you show that the call is welcome. Think about who are person is, how the person feels, and what the person is trying to accomplish. Create a mental picture of the person and take the time to listen to what the person has to say. If you do all these things, you will find it easy to practice the most successful technique of human relations—you will find a good reason to really like the person with whom you are dealing.

Once your voice is properly directed toward the transmitter, your attention is properly directed toward the call, and you have established a genuine appreciation for the caller, your good business personality will travel across the wires like a winning smile.

The telephone set in the background is a Stowaway, one of AT&T's Design Line Telephones. Photo reproduced with permission of AT&T.

■ ANSWERING THE TELEPHONE

Often the first thing you must do when the telephone rings is to accept the interruption. You must stop whatever you are doing. Above all, you must stop your present conversation before you open the line, especially if you are angry or laughing. Hearing the end of any conversation is disconcerting to even the most understanding caller. Never assume you know who is calling, even if someone intended to "call you right back" or if you expect a call with a great deal of certainty. Assumptive answering can most assuredly get you into embarrassing situations.

If you have a multiline telephone, look before you leap. Check your instrument to be sure the button is depressed for the line that is ringing so you do not break in on someone else's call. If you pick up a line on which someone is dialing, that person will be unable to make a connection and will have to start the call again.

Pick up your receiver only when you are certain that both you and your telephone are ready to answer the call. This should be sometime during the second ring. The caller is thrown off balance if you answer sooner because he or she expects to hear at least one full ring before there is an answer. Inefficiency is suspected after three or four rings on a business phone. Worst of all, business can be lost if calls go unanswered, and business callers frequently give up and assume the office is closed after five or six rings.

What you actually say when you answer the phone will be determined primarily by company policy, but sometimes by your own preference. The standard patterns for answering a business phone may include the following elements.

☐ Greeting

Alexander Graham Bell was disappointed because people did not answer the phone by saying "hoy hoy." Perhaps it is just as well, however, since *hoy* originated as an expression for driving animals! *Hello* originated as an expression of surprise, but it is more appropriate. Hello is accepted in business offices, but the preferred greetings are the more formal good morning, good afternoon, and good evening. Keep in mind that the greeting does more than just start the conversation on a cheery note. It also allows the caller a very important instant in which to stop waiting and start listening.

☐ Identification

Identification is where actual communication begins. It tells the caller whether or not the right connection has been made, and it must be done very clearly and precisely. If you allow habit to cause you to rush through your statement of who you are, you will garble the words and begin your conversations with miscommunication. Your identification may include one or several of the following: name of the company or your department; your title or function; your sirtitle and last name, your full name, or just your first name. Given next are examples illustrating how your position in the firm will determine the means you use to identify yourself.

If you are a console attendant, you need only identify the company:

Good morning, Sims Products.

If you are an order clerk and your calls are routed to you through a switchboard, your personal identity becomes more important:

Order Department, Ruth Stewart speaking.

If you are an executive and your calls are screened, your identification can be brief, but not abrupt:

This is Mr. Sims.

How's your telephone etiquette? Ever realize that you can be one of the top public relations people in your company? Well, you can! When someone calls you on the telephone, as far as he/she is concerned, you are the company. When you talk, it's the company talking. And if you sound cheerful, clear and interested, the caller's attitude toward the company will be a good one.

Below are a few helpful "do's and don'ts." After reading them carefully, if you decide your telephone etiquette is perfect, congratulations. If not, why not try to improve? Your telephone etiquette, after all, is sometimes the only way people can judge you — and their favorable judgment can mean dollars and cents to you and your company. So let's talk business.

See yourself as others hear you Ever stop to wonder how you'd sound if you could call yourself? You'd find your speech has four important characteristics: cheerfulness, distinctness, volume and speed.

Grab that phone Nobody likes to be kept waiting — especially on the telephone. So answer yours promptly. If you can't, explain the delay and apologize for it. Your caller may have good news or an important message and may not have time to wait.

Many telephone companies offer excellent leaflets free of charge on good telephone etiquette, such as this one published by GTE. Reproduced with permission of General Telephone Company of California.

Don't trust to memory If there's a message, write it down. Even if there's no message, make a note of who called and at what time. If the caller would like the call returned, be sure to get the name and number accurately. Your handling of telephone messages is a mark of your efficiency.

Wrong number Nobody enjoys answering or calling wrong numbers, and there's a very simple way to avoid them. Pronounce the number you want distinctly. Maintain your own frequently called number list.
Always look up doubtful numbers in your directory and dial carefully.

If answering party sounds unfamiliar, ask pleasantly,"Is this 555-9999?" If not, apologize briefly and re-check your directory.

When answering a wrong number call, refrain from slamming the receiver with a curt "Wrong number!" Instead say, "Sorry, there's no Bill Evans here." This is not only more courteous, but you won't get another call asking for Bill Evans.

It's all in the way you call There is always the chance that the person you're calling may be very busy. So to save time identify yourself quickly and state your business. Be cordial, but be informative and businesslike. Incidently, never try to fool an operator or a secretary into thinking you're a personal friend of Mr./Ms. Whatzit if you're not. You may get away with it once, but the next time you call, watch out!

What was that crack You wouldn't think of slamming the door when a visitor leaves your home. Give your telephone caller the same courtesy. At the end of the call, hang up gently. Don't bang the receiver so your caller gets a sharp crack in his/her ear.

Don't kick a caller around Nothing is more irritating than "the telephone runaround" — being kicked from one extension to another. When you get a call, handle it if you can. If you can't, tell the caller you'll transfer him/her to the right party. Then do it — with all the promptness and care you would expect of someone else. It's worthwhile treating all calls as important. Most of them are!

Be easy to trace

When you leave your office for any length of time, don't keep your absence a secret. Someone may call for information which only you can give. So let it be known where you can be reached, how long you'll be there, and when you'll be back. That way, if someone calls, whoever answers won't have to embark on an office-to-office manhunt. Time is money, so when you leave, leave word.

And talk normally. Some people — you probably know one or two — try to change their personalities when they pick up a receiver. Some roar like a bull moose paging its mate. Others whisper as though their message were a deep dark secret. Still others try to sound sophisticated...or mechanical...or cute.

Be cheerful. It will make your caller feel good and you'll feel better too. If you're able to get that "glad to hear from you" ring in your voice, you'll find callers becoming just as pleasant.

The best telephone voice is your own. So be natural. It's easier, simpler, better business all around.

Tell the world who you are

If you want to save time, and sound businesslike too, tell your caller right off the bat who you are. "Sales Department, John Brown," sounds efficient, saves your time and your caller's. "Yes" or "hello" means nothing. If you answer someone else's telephone, "Ms. Gaffney's office, Jim Williams."

Find out who

If you answer someone else's phone and the caller doesn't identify himself/herself, try to find out tactfully. Don't, of course bark "Who are you?" It may be the Chairman of the Board, and you'll find out! A better way is to ask, "May I tell him/her who's calling, please?" or "May I have him/her call you?" And don't put down the telephone until you're sure the conversation is over — the caller may want to talk to someone else.

People are proud

Most people are like the rest of us — in love with their own names and titles. So where you can, use them frequently. "Professor," "Colonel," "Doctor," — even the simple "Mr." or "Ms.", followed by the name, is sweet music to their ears.

Try to interweave names generously with other expressions of courtesy and respect. "Thank You." "Sorry." "Your're Welcome."

Unless you're calling to foreclose the mortgage or borrow fifty dollars, you can't miss having your call remembered pleasantly. And pleasant association — goodwill — is the cornerstone of good business. That's important to you, too, because you can be one of the top public relations people in your company.

GTE

Telephones of some professional people may be identified with, for example, "Doctor's Office" or sometimes with the telephone number only. Whatever means of identification you use, however, be sure you tell the caller who you are and if the right connection has been made.

In a busy emergency ward of a large medical center, the most reassuring word that a caller can hear is:

Emergency.

☐ Service Question

The commonly used service question is "May I help you?" Like the greeting and the identification, the service question can be more than just a polite gesture; it can perform an important function. Since business calls are frequently routed through one or more intermediaries, the service question can tell the caller when to wait for the call to be routed and when to begin the conversation. Time and patience can be saved if the caller does not have to state the purpose of the call several times before reaching the correct destination.

■ HANDLING CALLS

Some of the services performed for callers involve the mere handling of the call itself. These services include holding calls for available lines, routing or transferring calls, screening calls, or taking messages. Though they may seem simple and even trivial, such services must be performed with expressed consideration for the caller. Because the caller cannot see what is being done, even the slightest abruptness can be misinterpreted and good will can be lost. Remember that the caller will measure the company and its product by the efficiency with which calls are handled.

☐ Leaving the Line

No matter what your position may be in a firm, you will frequently have to place a caller on hold while you leave the line. Perhaps you must obtain information from another location or you must ask a visitor to leave your office during a call that is confidential. Many telephone systems require you to leave the line in order to transfer the call to another office. Whatever the reason, you can do so with maximum consideration for the caller by following these steps:

1 Tell the caller what you are going to do.

Excuse me. I am going to place you on hold while I close my door.

or

Just a moment, please. I must leave the line to get the information from another office.

or

If you will hold the line, I will transfer your call to the department that can help you.

2 Allow the caller to respond. True, many callers do not respond. However you must not risk cutting off any response, no matter how insignificant it may seem.

3 Depress the hold button (if you have a key telephone). It is so easy to forget! Relax, however. You need not have irrational fears about losing the connection. If you get back on the line within a second or two you can save the call. If you do lose the connection, you have not killed the customer! The originator of the call should simply place the call again.

4 Return to the line every 30 seconds, if possible. Not even a desert island is as lonely as a telephone line on hold. A simple "still checking" or "still holding" is reassuring. Again, allow the caller to respond each time before depressing the hold button. If it appears there will be a long delay, offer to call back.

5 Express appreciation when you return. "Thank you for waiting" tells the caller that you are back and your conversation will resume. It also provides that all-important opportunity for the caller to stop waiting and start listening.

□ Routing Calls

If you are a switchboard operator, console attendant, receptionist, or secretary, you may route hundreds of calls every day. In any other position, you will find yourself answering the phone and routing calls from time to time, or perhaps rerouting a call that has been misdirected to you.

There are three aspects of routing calls within the organization that must be understood by everyone in the office: (1) how to operate the equipment, (2) what to say, and (3) how to decide what person or department should receive the call. Equipment is covered in an earlier chapter, words and responses are discussed here, and practice for the decision-making process is provided at the end of this chapter.

It is important to note at this point, however, that the caller often does not ask for a person or a department by name. The routing decision may have to be based on the caller's statement or request, which is sometimes inadequate. Therefore, everyone in the office must understand the organization of the company and the functions of various departments and people. Such an understanding starts with a thorough study of the organization chart and company manual. All the examples and exercises used in this book are based on the fictitious company, whose organization chart is given in the figure on the facing page.

Choosing your words and responses will be the easiest part of the task. Of course, you will not use exactly the same words every time, but the patterns below will always work smoothly for you.

1 An easy call

You: Good morning, Sims Products, May I help you?

Caller: May I speak to Mr. Jackson, please?

You: Yes, just a moment and I'll connect you.

or

Yes, but he's on another line right now. Would you like to hold?

2 A decision call

You: Good morning, Sims Products. May I help you?

Caller: I'd like to speak to someone about my bill.

You: Yes, Marjorie Arnoff in Accounts Receivable can help you. Just a moment and I'll connect you.

3 A redirected call

You: This is the Accounting Department. May I help you?

Caller: I'd like some information about your frypans.

You: Yes, our Appliance Sales Department can help you. Just a moment and I'll transfer you to Ms. Green in that department.

Notice that each time you responded to the caller's request, you began by saying "Yes" and followed by explaining what you were going to do. These are the simple words that inspire the confidence and good will of the caller, which is vital to every call, whether it is as routine as these or as delicate as some that will be illustrated later. Finally, take special note of the fact that it was not necessary in the redirected call to refer to a mistake or blame someone for an error.

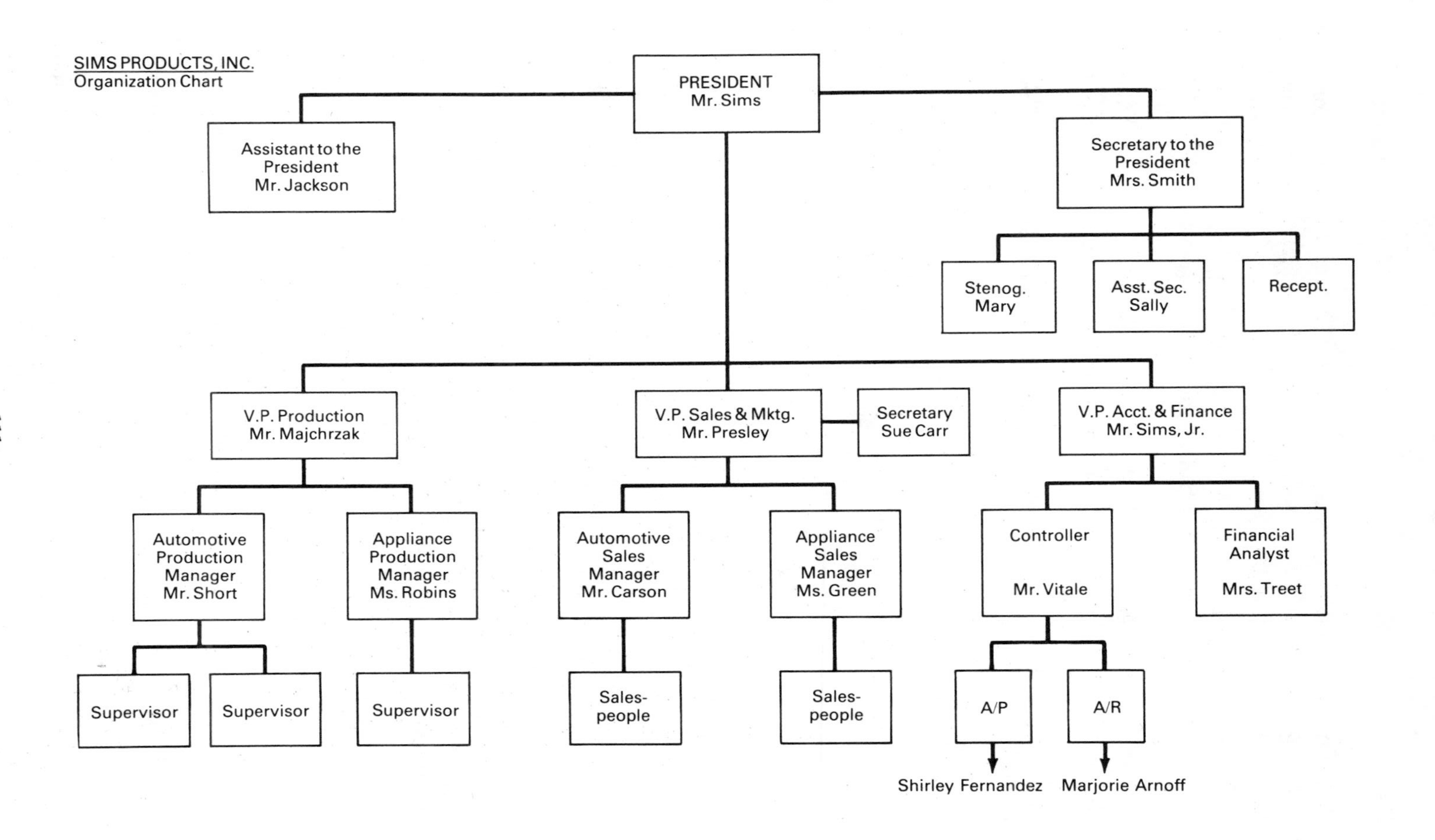
SIMS PRODUCTS, INC.
Organization Chart
PRESIDENT
Mr. Sims
Assistant to the President
Mr. Jackson
Secretary to the President
Mrs. Smith
Stenog.
Mary
Asst. Sec.
Sally
Recept.
V.P. Production
Mr. Majchrzak
V.P. Sales & Mktg.
Mr. Presley
Secretary
Sue Carr
V.P. Acct. & Finance
Mr. Sims, Jr.
Automotive Production Manager
Mr. Short
Appliance Production Manager
Ms. Robins
Automotive Sales Manager
Mr. Carson
Appliance Sales Manager
Ms. Green
Controller
Mr. Vitale
Financial Analyst
Mrs. Treet
Supervisor
Supervisor
Supervisor
Sales-people
Sales-people
A/P
A/R
Shirley Fernandez
Marjorie Arnoff

Final Reminder: Your next steps are to *allow the caller to respond* and *depress the hold button* (on a key instrument) before routing the call to its final destination.

☐ Announcing Calls

The call will reach its final destination when you either signal the recipient that a call is waiting or call the recipient on the intercom line to announce the call. The signal can be transmitted by either pressing a button connected to an audible signal on the recipient's desk, pressing the button for that station in a direct station selection field, or dialing the station on the intercom path. You announce the call on the intercom by telling the recipient something like the following:

> There is a call for you on line 1.

or

> Jack Smith is on the hot line.

or

> Your daughter is calling long distance on 4.

or the extra respectful

> Will you talk to Harry Turner on 1675?

When you announce a call that has been misdirected to you and that your are transferring to another person, tell the recipient the nature of the call so the caller does not have to repeat his or her request. This saves everyone's time and is especially helpful to the caller.

■ SCREENING CALLS

Contrary to public opinion, a screened telephone should be neither a status symbol nor a technique used by snobs. Screening provides important administrative support to busy people who receive so many calls or have so much responsibility that some control must be exercised over interruptions. In addition, some executives may delegate certain areas of responsibility to subordinates, and incoming calls must therefore be carefully directed. Three types of screens are illustrated here, each of which serves a different purpose.

☐ The Information Screen

Some office workers transact so much business over the phone that they can work more efficiently if they know who is calling and what the call is about before they answer. The information screen provides an opportunity for preparation—possibly to locate a file, to clear the office of visitors, or simply to recollect the caller or the issue that may be prompting the call. Here is an effective pattern for such an information screen. Either or both questions may be used.

You: Good morning, Mr. Sim's office.
Caller: I'd like to speak to him please.
You: Yes, may I tell him who is calling?
Caller: This is Mr. Jones, with the Chamber of Commerce.
You: Thank you, Mr. Jones. Will he know what your call is about?
Caller: It is about next month's meeting.
You: Just a moment and I will connect you.

Notice that your first responses was "Yes," which inspired immediate confidence and told Mr. Jones he would succeed. Then you set up the screen with a question that showed you wanted to facilitate the call rather than block it. With few exceptions, this question will elicit the response you want. Most callers will not hesitate to give you their names, the names of their companies, and the natures of their calls. Notice also that you repeated Mr. Jones' name in your response to him. You did this for three important reasons: (1) to be sure you heard his name correctly, (2) to help you remember it when you tell Mr. Sims who is calling, and (3) because names are important to their owners. People like to hear their names, and they like their names to be pronounced correctly.

☐ The Partial Screen

Some administrators are so busy that they must work with priority schedules and cannot always be available to talk to every caller on the spot. They absolutely must know who is calling before they decide whether or not to take the call. If it is your job to control a phone in such a busy office, you can use the screening technique shown in this pattern:

> **You:** Good morning, Appliance Sales Department.
>
> **Caller:** May I speak to the manager please?
>
> **You:** I'm sorry, Ms. Green is not available right now. May I ask who is calling?

You use this response for the partial screen regardless of where Ms. Green is or what she is doing. She may be in a meeting on the other side of town, or she may be standing right next to your desk.

Study the details of that screen carefully. You began with "I'm sorry," indicating your sensitivity to the position of the caller in the screen. Then you established the screen without making excuses for Ms. Green that may have to be coordinated with her later. Finally, you politely asked for information so that the caller could know you wanted to help. With your good service attitude, the call will most likely proceed as follows:

> **Caller:** Yes. This is Mrs. Armstrong from the accounting department in the Newark office.

At this point, you must exercise your judgment about whether or not Ms. Green will want to answer the call. Your screen is called a *partial screen* because some of the calls will get through. You are prepared through instruction and experience to decide which calls to route and which calls to block. If you decide Ms. Green would not want to take this call right now, you can maintain the screen quite easily.

> **You:** She is not at her desk right now. May I have your telephone number and I will ask her to return your call?

Study this response carefully. You have paved the way for the call to occur at Ms. Green's convenience, and you have done so without committing her, without making specific explanations or excuses, and without offending Mrs. Armstrong. Notice especially that you said, "I will *ask her* to return your call." You could also have said, "I'll see that she receives your message." Under no circumstances should you say, "I'l have her call you," or "she will call you this afternoon." You can tell the caller what *you* will do, but you cannot tell the caller what Ms. Green will do.

Now let's go back to the point where you exercise your judgment and assume that you decide Ms. Green will want to talk to Mrs. Armstrong right away. You can break the partial screen quite smoothly by saying:

You: Oh yes, Mrs. Armstrong. Just a moment and I'll try to locate her for you.

Now you must announce the call to Ms. Green by calling her on the intercom line and telling her that Mrs. Armstrong is waiting. If your decision turns out to be incorrect and Ms. Green does not want to talk to Mrs. Armstrong, you can return to your screen easily:

You: I'm sorry, Mrs. Armstrong, but I'll have to ask her to return your call.

or

Can she call you back this afternoon? She is really swamped right now.

or

Is there a number where she can reach you later? She just can't get to the phone right now.

Most executives who employ the partial screen or any variation of it intend to respond to every call. The screen merely enables them to control their workflows for maximum efficiency and to do a better job for everyone.

☐ The Full Screen

Some executives delegate very large areas of responsibility and authority to subordinates and answer practically no incoming calls at all. They may even have separate telephone lines for their own use. In such a case, it is necessary for incoming calls on the regular line to be controlled by a full screen.

Consider, for example, the district attorney for a large metropolitan area, the claims manager of an insurance company, or the vice-president in charge of marketing for a multinational corporation with a variety of consumer products. The following patterns can be used to control and direct calls to the right person, who can respond to the caller's need immediately. In the full screens illustrated, you will see how important it is for the person answering the telephone to understand the organization of the firm. You may want to refer back to the organization chart for Sims Products Company to understand the routing decisions. As you will see, even our own Mr. Sims delegates extensively to his vice presidents and managers.

Receptionist: Good morning, Sims Products. May I help you?

Caller: May I speak to the president, please?

Receptionist: I'm sorry, Mr. Sims is not available right now. Can someone else help you?

The receptionist starts out by showing sensitivity to the position of the caller in the screen. This sensitivity is important to the success of any screen. Then, the receptionist asks a question known as a *probe*. The objective of the probe is to find out something about the purpose of the call and then make the proper routing decision.

Caller: I want to talk to him about some trouble I'm having with my toaster.

If the probe works this well on the first try, the receptionist can immediately route the call to the department that can do the best job in handling the problem.

Receptionist: I'm sure Ms. Green, our Appliance Sales Manager, can help you right away. Just a moment and I'll connect you.

Of course, the caller may be quite determined to talk to the top person, and another call might go like this:

Receptionist: I'm sorry, Mr. Sims is not available right now. May I ask what your call is about?

Caller: No! I really must talk to your president about this.

The skilled receptionist will try another probe. This time, an open-ended question may be used in an effort to get the caller to reveal the nature of the call. Any scrap of information revealed can be used as a wedge to make the decision and route the call.

Receptionist: What sort of problem are you having?

Very few callers are insistent beyond this point, because they really want to tell the story immediately. It takes just a little polite encouragement to get them started and then connected with someone who can take appropriate action.

Of course, there are those "I-must-go-to-the-top" callers who put even the most skilled receptionist to the test.

> **Caller:** Well, I need to talk to him right away. You people shipped us 3000 of these things, and they don't fit our product. What's more, we only ordered 300. I want him to know about this kind of efficiency.

But watch this skill as the screen is maintained.

> **Receptionist:** He certainly will want to know about it, Mr. ah . . . ah . . . . May I take your name and number, and I will ask him to return your call?
>
> **Caller:** Yes. My name is Robert Miller, and my number is 555-7643. I'm with Interstate Auto in Dallas.

The call is then routed immediately.

> **Receptionist:** Oh, yes, Mr. Miller. Well, Mr. Sims won't be available until late this afternoon. I'll see that he gets your message. In the meantime, Mr. Carson, our Automotive Sales Manager, could help you right away. Would you like me to connect you with him?

Will Mr. Sims return this call? Yes, he probably will. First, however, he will consult Mr. Carson to get all the facts. Then, if he feels he can help build good will, he most assuredly will telephone Mr. Miller. Was the screen successful? Yes, it was. The purpose of the screen is not to keep Mr. Miller from talking to Mr. Sims, nor is it to keep Mr. Sims from finding out about things that go wrong. The purpose of the screen is to connect Mr. Miller with Mr. Carson for immediate action, while permitting Mr. Sims to retain control over his incoming calls. The receptionist here has made a vital contribution to the efficiency and success of the company.

▪ TAKING MESSAGES

You undoubtably observed that several of the calls illustrated on the preceding pages resulted in messages. As you learned earlier, only one out of every four business calls reaches its destination on the first attempt because it is difficult for business people to be available at their desks all the time. Therefore, messages are an accepted part of the communication process.

Some telephone messages in offices are merely scribbled notes on scraps of paper that are discarded after the call is returned. More often, however, messages are very important tools for accomplishing business tasks, and sometimes they even become part of a company's permanent files. They may be written in duplicate in a spiral-bound book, with the original sent to the recipient and the copy retained as a chronological record of telephone activity. The written message may even become an official record of a business transaction. Many business people refer to message records when they need to recall the date of a conversation, a name, a telephone number, or a fact that was part of the message.

Every telephone message should be written with great care. You should learn to take messages without rewriting them, for copying increases your chances for making errors. You can avoid rewriting by confirming or repeating each name, number, and message. In this way, you will not only be sure that you hear everything correctly, but you also allow yourself enough time to write everything completely and legibly.

The message form shown in the figure has been completed for the call from our very insistent Mr. Miller. Study it carefully and you will see that it has the six essential ingredients that should be included in every telephone message you write:

1. Date and time of the call.
2. Name of the person being called.
3. Identification of the caller, including title and company name.
4. Phone number of the caller, including area code and extension.
5. Message.
6. Identification of the person who took the call.

When Mr. Sims reads this message, the questions that will formulate in his mind are: Who is this person, and what does he expect of me? Each of the six items you wrote on the message form will help him answer those two questions.

The date and time become increasingly meaningful as time elapses between the call and the time Mr. Sims receives the message. For this situation, Mr. Sims will try to gauge the possibility that Mr. Carson has already resolved the problem. It is even possible that communication may occur for some other reason between the time the message is written and the time it is received. In this event, the time indicated on

Date 3/16/82 Hour 10:30 a
To Mr. Sims
WHILE YOU WERE OUT
M r Robert Miller
of Interstate Automotive

TELEPHONED	X	PLEASE CALL	X
CALLED TO SEE YOU		WILL CALL AGAIN	
WANTS TO SEE YOU		RUSH	
RETURNED YOUR CALL			

AREA CODE NUMBER EXTENSION
Phone 900 555-7634
Message about a problem with their order for auto parts; referred to Carson
ss
SIGNED

DATE HOUR
WHILE YOU WERE OUT
ONE AREA CODE NUMBER EXTENSION

TELEPHONED		RETURNED CALL		LEFT PACKAGE	
PLEASE CALL		WAS IN		PLEASE SEE ME	
WILL CALL AGAIN		WILL RETURN		IMPORTANT	

MESSAGE
NED

NAME Ramadan, Ms
ADDRESS 3916 Main Street DATE 10-1-82
ITY Anytown STATE USA ZIP 00000
ONE 900-555-8911 SOURCE JT
ES to SP
APPOINTMENT 10-4 2:30 p
CAT BROCH ✓ LEAD O-SEAS EXEC MED
LEGAL ACCT ✓ BZ MGT ✓ F AID ✓ MAILED 10-1

WHILE YOU W
TO: HOU
DATE
MR./MS
OF:
☐ Telephoned ☐ Please Call ☐ Will Call Again ☐ Returned
☐ Was In To See You ☐ Will Be In Again ☐ Please See Me ☐ Imp
MESSAGE:
Signed

the message form tells Mr. Sims whether this is a new call or a call that has already been answered.

The company name is especially important in helping Mr. Sims identify the caller and the nature of the call. The title and company *combined* usually provide substantial clues about the nature of the call, even if the caller is not known and there is no message. Correct spelling of both the caller's name and the company name is absolutely essential at this point. Do not assume that Mr. Sims will know how to spell a name or that you will have an opportunity to check it later. Spell it while you are taking the message and confirm it. Remember, confirming also serves the added purpose of giving you time to write the message clearly and legibly.

The importance of a complete and accurate telephone number cannot be overemphasized. Because so much business is transacted over the phone, the number is almost as important as the last name! Include the area code, even if it is the same as yours. If you omit it, the recipient may not know whether the number is local or you forgot to ask for the area code. Always ask if there is an extension. *Confirm and repeat every number.* If you do not pay very close attention, you can transpose the numbers, write a seven that looks like a one or a nine, or—worst of all—omit one of the digits altogether. Giving someone a telephone message with an inaccurate number is like giving someone half of a $100 bill!

The actual message may be as simple as "Mr. Jones called," or it may be a very complicated set of instructions. The forms illustrated allow you to check off some of the more common messages. Additional information may require considerable discretion on your part. If the caller wants to transmit some lengthy or complicated information, you may be required to paraphrase and abbreviate. The message from Mr. Miller illustrates how such information can be kept brief, but complete.

The last ingredient, your identification as the person who took the call, is far more important than it may appear. The recipient may want to ask you, "How did she sound?" or "Did he say where he was going?" or "Do you know what time their office closes?" It is very probable that you can help. Finally, if you do make an error, you are the person with the best chance for correcting it.

The other filled-in message form on page 119 deserves special note. This company uses its telephone messages to make statistical analyses, and each message becomes part of a permanent file. To use this form properly, you would need special instructions from your supervisor about what information is requested and the meaning of each abbreviation.

▪ PLACING CALLS

When you make a call from an office, the telephone is an important tool to help you perform a specific task. The telephone merely enables you to go somewhere. Once you are there, it is up to you to get the job done. Your objective is to complete the task with one call. Inadequate preparation for a telephone call can cause more than mere delay; it means the call must be placed again. Misconnections and incomplete calls can be costly and are frustrating for both you and the people you may interrupt. Careful preparation enables you to use your telephone efficiently and do your job effectively.

□ Planning the Call

Most of the advance preparation will involve planning the conversation and your approach to the task at hand. Here are some questions you will want to consider beforehand:

1. Whom are you calling? If it is someone you know, call up a mental picture of that person. If it is someone you do not know, try to anticipate what that person may be like. You can practice good human relations by showing that you have an understanding of the individual to whom you talk. If you do not know exactly who can help you, be sure you have formulated a clear statement of your request or purpose so that you reach the right person in the shortest possible time.
2. Is this the best time to call? What are the chances that the person you are calling will be available? Is this a time he or she is likely to be extremely busy, in a regularly scheduled meeting, or attending an important convention? Is this the time he or she is likely to be receptive to your call or purpose? If you are making a long-distance call, you must know the time at your destination. Special considerations for calling across time zones and international date lines are covered in Chapter 4.
3. Should you make some notes on the specific questions you want to ask and the specific points you want to make? What is the best order in which to ask or make them?
4. What information, facts, files, or records should be handy for you to accomplish all your objectives? What questions are you likely to be asked during the conversation?
5. What is the complete phone number, including the area code and extension? Can you find it on a phone message, in a personal directory, in your company directory, or in the telephone book? For information on directories and information services, see Chapter 4.
6. How can you minimize the cost of the call? Does your firm have special lines or services for certain geographic areas? Long-distance services are also covered in Chapter 4.

□ Connecting and Misconnecting

Even with careful planning before you pick up the receiver, you still may be among the three calls out of four that do not reach their destinations on the first try. Here are some of the "detours" you may encounter.

These Rolodex ™ directory files provide maximum flexibility and efficiency. Photos courtesy of Rolodex Corporation.

1 The person you call may be unavailable. In a business office, however, you are likely to reach a receptionist, operator, or secretary so that you can leave a message. On some occasions, you may get a busy signal or no answer and have to try again later.

2. You may reach the wrong destination, either because you dialed incorrectly or because you did not have the right number. Occasionally, the phone company's switching system may cause the problem. You can determine the cause of the error by asking what number you reached. The response will tell you exactly what went wrong. In case you reach a person who does not wish to give out the number, it is polite to say, "Have I reached 555-2354?" Whatever the problem may be, you can report your error to the operator and get credit if it is a toll call. If you do not have the right number, consult your files or telephone directories before requesting operator assistance.
3. You may get a recording. Most recordings are self-explanatory and help you accomplish your task. You are probably familiar with the recordings that tell you a number has been changed or is out of service. A recording you encounter at your destination may tell you that all the lines are busy and you will be connected as soon as someone is available or that no one is there and you can leave a message on an automatic answering device, or it may even attempt to tell you the information you want and refer you to another number for additional information.
4. You may encounter a screen, such as those described earlier in this chapter. Remember, it is best for you to cooperate because the screen is probably designed to facilitate your call. However, even a full screen can be broken if you are willing to make a second call. Simply ask the secretary when the person you wish to reach will be available and say you will call back then. In effect, you have arranged an appointment for your call, and you are likely to succeed.
5. On rare occasions, you may dial correctly and get no ring at all because all the circuits somewhere between you and your destination are busy. Usually, after a delay, you will get a very rapid busy signal (120 impulses per minute, as opposed to the 60 impulses per minute indicating that the station is busy). If you reach a station that is out of order, you may hear a ring but the person you are calling does not. Sometimes, even with modern switching systems, you may get no connection at all.

In any of these events, place the call again immediately. If you do not get a connection the second time, request operator assistance.

One final and important note: Long ago, when all long-distance calls required operator assistance and connections took a lot of time to make, it helped busy executives to have calls placed by secretaries or assistants. Today, it is recommended by all telephone companies and is nearly always more efficient for everyone to place his or her own calls. Delays are usually caused when people are not available, not when circuits are busy. In such cases, it is faster and better for the person placing the call to determine the next step first-hand, rather than by a three-way conversation through an assistant. With modern, automatic on-hook dialing, time spent placing calls need not be wasted.

A LIST OF USEFUL RESPONSES

Answer ring	Good morning, Sims Products, Miss Smith speaking.
	Accounting Department, this is Sally.
	Mr. Sims speaking.
Service question	May I help you?
When the person called is in	Yes, just a moment and I'll connect you.
When the person called is on another line	Yes, but he is on another line. Would you care to hold?
When the person called is not in	I'm sorry, she is not in the office right now. Would you like to leave a message?
	He's away from his desk, May I ask him to call you?
	Ms. Green isn't available right now. Can someone else help you?
Information screen	Yes, may I say who is calling, please?
	May I tell her what your call is about?
	Will he know what your call is about?
Partial screen	I'm sorry, she is not available right now. May I ask who is calling?
Break partial screen	Oh, yes. Just a moment and I'll try to locate her for you.
Maintain partial screen	She is not at her desk. May I ask her to return your call?
Full screen	I'm sorry, he is not available. Can someone else help you?
Probe	May I ask what your call is about?
Transfer the call	Mr. Jones is in charge of that department, and I'm sure he can help you. Just a moment and I will transfer your call.
	Our Marketing Department can answer your question. Just a moment and I'll have your call transferred.
Closing	That answers all my questions, and I thank you for all your help.
	This is all the information I need.
	It was nice talking with you again.
	If you have no more questions, I'll look forward to hearing from you when you are ready to order.
	I have another call waiting, and I must hang up now.

	I know you are busy and I don't want to keep you.
	Good-bye.
Additional explanations	Mr. Sims will be out of town until next Wednesday, but I expect to hear from him this afternoon. May I give him a message?
	Mrs. Armstrong will not be in today. Can someone else help you, or would you like to leave a message?
	Dr. Adams is taking calls for Dr. Ellis. May I give you that number?
	Ms. Miller is on vacation. May I connect you with her secretary?
	That department is closed for the day. Their hours are from 8 to 4:30. Would you care to call back tomorrow?
	That department is located in another building. Just a moment and I can patch you through. The number there is 555-4230 if you want to call them in the future.

■ CLOSING THE CALL

Whether you place or receive the call, you can sense when the call is complete and the task is done. Either party to the call can signal the end of the call with an expression such as:

> Well, that really answers all my questions. I appreciate your help, Susan.

or

> It was nice talking with you again, Jack. I'll look forward to seeing you at next month's meeting.

or

> Thank you for calling. If we can help you again, please let us know.

or

> I know you are busy and I don't want to keep you. Thank you for all your help.

Or you may feel you need something stronger than a signal, and you wish to be as direct as:

I have some people waiting.

or

I have a call coming in on another line.

or

We're really swamped here today, and I've got to go now.

Your closing should always show polite consideration for the caller and end with good-bye. Then, no matter how pressed you may be, allow the person to respond before you hang up the receiver, and take care to replace the receiver gently. Whether your telephone call involves routine tasks or high-level contract negotiations, your respect for the dignity of the other person is the key to successful communication, just as it is vital to a successfully managed business.

QUESTIONS FOR REINFORCEMENT AND DISCUSSION

1. How soon should you answer the telephone?
2. Name three elements that might be included in what you say when you answer the telephone.
3. Why is it important for you to repeat the caller's name? (Give *three* reasons.)
4. What are the two things you must remember to do before leaving the line (assume a key telephone)? How often should you return to the caller on hold?
5. How can the company's organization chart be of value to the person routing telephone calls?
6. Describe some ways in which you can achieve the right attitude for successful telephone conversations.
7. Describe the kind of administrative support provided by each of the three types of telephone screens.
8. What are the six elements in a complete message?
9. Why is the time of the telephone call an important part of the message?
10. What are two reasons for confirming spellings and telephone numbers when writing messages?
11. Why is it important for the person taking the message to be identified on the message form?
12. What are some considerations for planning a telephone conversation?
13. Describe a conversation using good technique for a call that has reached a wrong number.
14. What are some expressions you can use to close a call politely (before you can say good-bye)?

15. Assume you are a sales correspondent for a large mail-order firm. You receive a call from a customer inquiring about an error on her bill. Below are three actions you might take, and all three would get the job done. Tell which of the three is best, which is second best, and which would be the least efficient, and why.
    a. Transfer the call to the sales correspondent who handled the sale.
    b. Try to handle the call yourself by telling the customer you will check out the transaction and call her back.
    c. Transfer the call to the Billing Department.

TELEPHONE CALLS AND THE ORGANIZATION CHART

Study the Organization Chart for our fictitious firm, Sims Products, so you are familiar with each department and person and each one s functions and responsibilities. Then decide who you think should receive each of the following calls.

1. I want some information about your frypans.
2. I'd like to talk to someone about my bill, please.
3. May I speak to your legal counsel please?
4. Lemme talkta that guy Majorette.
5. I got this toaster from you people and it burned up. What are you going to do about that?
6. To whom do I speak about gear pinions for my automatic transmission?
7. Who is in charge of your accounts payable?
8. Do you have any openings for auto mechanics?
9. Who buys your office supplies?
10. I wanna talk to that lady upstairs in charge of your financial stuff.

ROLE PLAYING

Now that you are familiar with the organization chart for Sims Products, assume you are the receptionist and respond to the following callers. Mr. Sims' telephone is fully screened, and Mr. Jackson likes a partial screen when he is in.

1. May I speak to Mr. Jackson, please? (He is in.)
2. May I speak to Mr. Sims, please? (He is in.)
3. This is Mrs. Sims. May I speak to Mr. Sims, please? (He is down the hall rapping with someone about the all-star game.)
4. Hya, Sweetie. Put me through to Sims, willya? (He is playing golf.)
5. This is Mr. Crenshaw. I'm an attorney for the IRS. Is Mr. Jackson there? (He is in a meeting and asked not to be disturbed.)
6. Yea. Gimme Short.
7. May I speak to Mr. Rhodes?

8. Is my mommy there?
9. Is Junior in? (He is on another line.)
10. Lemme have Elvis Pelvis please.
11. May I speak to Dr. Carlin, please?
12. I wanna talk to that lady you got in your appliance department. (Ms. Greene is in a meeting in Mr. Sims' office.)
13. Who is in charge of your production department?
14. Sandy?
15. I need to know the model number on the cam shafts we ordered from you people last month.

PRACTICE TAKING MESSAGES

For practice, take the messages for the following calls as you read them. Use the current date and time.

First Call

You: Good morning, Sims Products. May I help you?

Caller: May I speak to Mr. Jackson, please?

You: I'm sorry, Mr. Jackson is not available right now. May I ask him to return your call?

Caller: Yes, this is Sam Boyle at General Motors. The number here is 555-8924.

You: That was 555-8924. Thank you. Mr. Boyle. Is your name spelled B-o-y-l-e?

Caller: Yes, that's correct.

You: I'll give him the message. Good-bye.

Caller: Good-bye.

Second Call

You: Good morning, Mr. Sims office.

Caller: Is he in?

You: I'm sorry, he is not available. May I ask him to call you?

Caller: Yes, this is Mr. Miyamoto. M-i-y-a-m-o-t-o.

You: And your company, Mr. Miyamoto?

Caller: ABC Electronics, 555-9862.

You: That's 555-9862. Is there a message?

Caller: Yes, tell him I'll be in Springfield Tuesday if he needs anything.

You: I'll see that he gets your message. Good-bye.

Caller: Thank you. Good-bye.

7 Communicating with People in Business

Now that you know how to reach your destination on the telephone let's get down to business. What you really want to do is communicate with people in business. In the past several decades, a great deal of emphasis has been placed on the importance of communication to the success of both the employee and the business. You will find that importance entirely justified in your career as a knowledge worker.

A good working knowledge of how to communicate begins with a clear understanding of the theory of human communication. An illustrative model is shown in the figure on page 130. Note the similarities between this model and our model for data communication in Chapter 5.

Simply speaking, *communication* is the transmission of an idea from one mind to another. If the idea received is the same as the idea sent, the communication is perfect. As you know from experience, however, most communication is just not that easy.

Perfect communication can be difficult to achieve, because the sender must express the idea in the form of symbols, a process we call *encoding*. Those symbols must then be interpreted by the receiver, a process we call *decoding* As illustrated in our model, the idea leaves the sender's mind, goes through a perception screen, and comes out as a

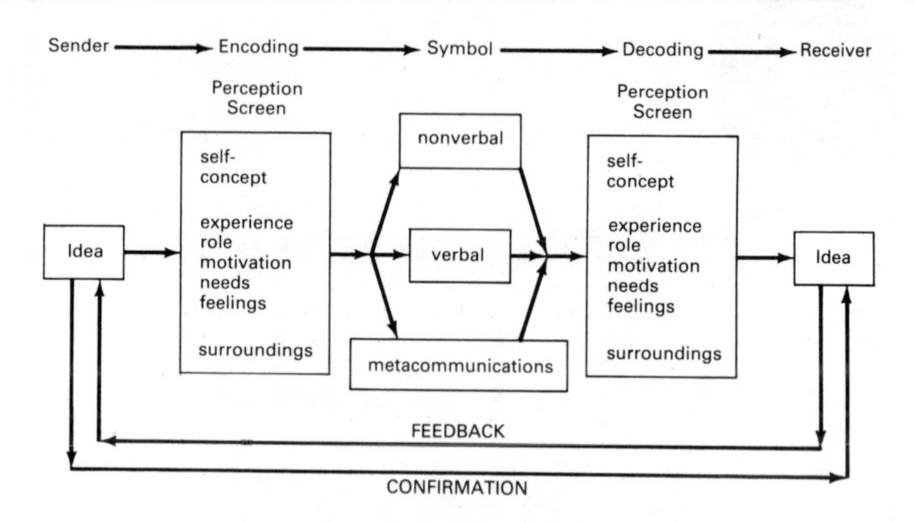

symbol. The symbol may be a word (verbal symbol), a gesture (nonverbal symbol), an implication (metacommunication), or any combination of these.

Each symbol then goes through the receiver's perception screen before it enters his or her mind as an idea. Imperfect communication can occur in either the encoding or decoding process because the sender and receiver may not perceive a symbol with exactly the same interpretation. In fact, human interaction can become so complex that it is nearly impossible for both the sender and the receiver to attach exactly the same meaning to every symbol. The resulting miscommunication can be a nuisance, a comedy, or a real disaster. Study illustrations shown on page 131.

Now let us study each component of the model to gain more insight into the complexities of human communication.

■ VERBAL SYMBOLS

Verbal symbols are words. We learn at a very early age to express our ideas with words quite easily and naturally. Then we study English in school to help us express ourselves more accurately and effectively. We may even learn other languages so that we can communicate with other peoples. No matter how expert we become in languages, however, we still experience some pitfalls with words as symbols of our ideas. There are several pitfalls you may recognize.

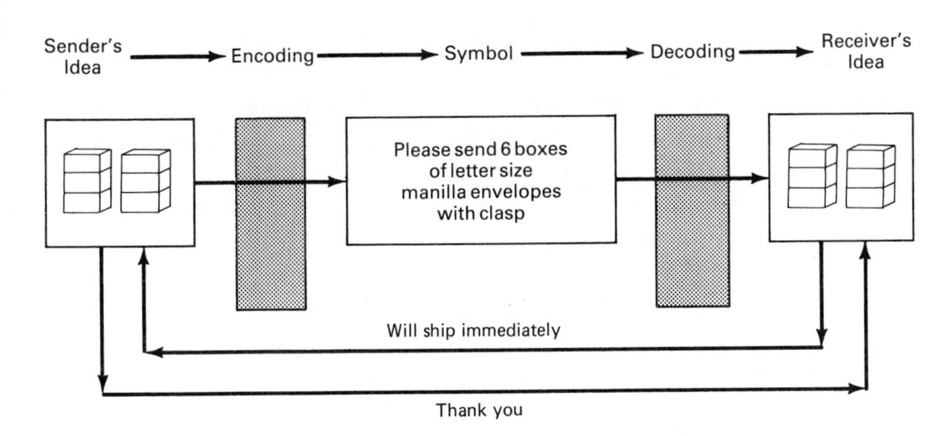

SUCCESSFUL COMMUNICATION

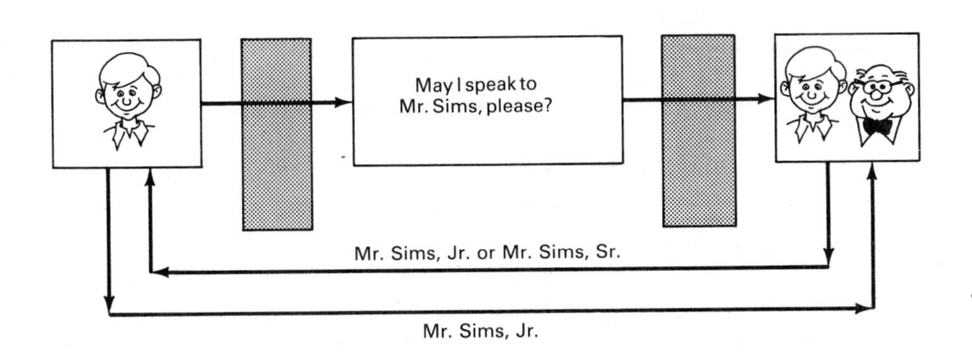

COMMUNICATION IN TROUBLE

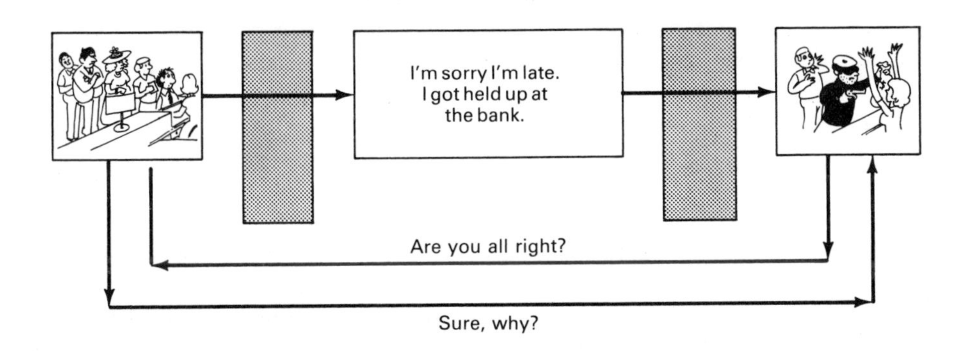

MISCOMMUNICATION

☐ Semantics

Problems of semantics may fall into one of two categories. First, some words have such broad meanings that exact interpretation may vary. For example, in the statment "this is a *serious* matter," the word *serious* may denote anything from "this is an important matter" to "this matter may have disastrous consequences." A second category of semantic problems arises with words that have several meanings. For example, the word *file* may denote any of the the following:

- A collection of documents pertaining to a certain subject, such as a company, person, product, or transaction.
- A folder in which documents are stored.
- A drawer, box, or cabinet in which folders are stored.
- A specific segment of information on tape, disk, or other media in a word processing or data processing machine.
- The act of storing documents in an organized system for easy reference or retrieval.
- A tool for smoothing a surface.
- To rub smooth, refine, or cut down with a tool or instrument.
- To march or proceed in a single line.
- To perform the first act of a lawsuit or criminal procedure.
- To register as a candidate for political office.

☐ Connotation

The definition of a word may be called its *denotation. Connotation* refers to the feelings that a word may elicit. Words with emotional connotation may have varying impacts on different people. For example, the word *profit* may have a very positive connotation to a stockholder but a negative connotation to a consumer advocate. Words such as *conserve, minimize*, and *skimp* have essentially the same meaning, but cause different feelings, as do *wealthy, rich,* and *fat-cat*. Also writers today avoid sexist connotations, such as referring to all telephone operators as "she" or to all managers as "he."

☐ Euphemisms

Euphemisms are "nice" words used to take the place of words with negative or unpleasant connotations. We refer to the rest room rather than a toilet, a supervisor rather than a boss, a senior citizen rather than an old person, and a maintenance person rather than a janitor; a person may be terminated rather than fired or departed rather than dead.

☐ Jargon

Jargon is great! Don't ever let anyone tell you it isn't. Any word or phrase that has special or technical meaning in a particular business, industry, or activity is *jargon*. It is rich with meaning. Its denotation is so specific that it conveys more exact ideas than any other terminology. For example, a stripper in a print shop masks a negative, while a stripper in a manufacturing process may have a certain painting task. A stripper in Las Vegas carries an entirely different denotation—as well as connotation!

Jargon has acquired a bad reputation for letter-writing purposes, but it is effective in conversations as long as you know you are talking with someone who speaks the same "jargonese" as you do.

☐ Verbosity

Some ideas may be easy to see but require entirely too many words to describe. To use a familiar cliché, a picture is worth a thousand words. Realistically, by the time you have waded through a thousand words, you may lose interest in the picture. A practical problem with verbosity occurs when you read the lengthy instructions for the operation of a machine and think it is the most difficult process you can imagine. Then you watch someone operate the machine and discover there is nothing to it. One of the biggest advantages of the Picturephone® and video conferencing is that they enable people on the telephone to see the subject of their conversation.

■ NONVERBAL SYMBOLS

Nonverbal symbols, such as actions and gestures, are commonly known as body language. Such symbols may be intentional or unintentional, and they communicate ideas and feelings quite accurately without any words. Recently, a great deal has been learned about the importance of body language in the communication process. Most nonverbal symbols must be seen, however, so they are lost in telephone conversations.

Among the nonverbal symbols discussed here, only voice quality can be perceived over the telephone.

☐ Facial Expression and Posture

A wink can say "I'm just kidding" or "You're special." A smile adds sincerity to "I'm pleased with your work." Arms folded across a chest say "I'm suspicious; you'll have to convince me." Head scratching or hair twisting say "I'm uneasy and nervous here." Slouching says "I'm bored with this," while an erect, forward posture says "I'm interested; let's get on with it."

☐ Actions

If you enter an office and the person behind the desk looks up and smiles, you know that person is saying "I'm glad to see you; what can I do for you today?" On the other hand, if that person continues writing or working, the message is clearly "I'm too busy to talk to you now; wait or come back later." Motioning toward a chair may say "Please sit down; my time is yours," while picking up a file folder or ledger says "I'm so busy that I must continue doing my work while you're here; don't stay too long." A person you are visiting in an office may tell you it is time to leave simply by picking up the telephone or standing up and beginning to walk toward the door.

☐ Space and Distance

Space and distance also carry messages. The amount of space occupied by someone's office or work station can tell a great deal about that person's status or position in the firm. The amount of distance maintained between you and another office worker may indicate your positions relative to each other. For example, to see a customer or your superior, you maintain some distance (or even stand in the doorway) until you are recognized. To see a colleague or subordinate, however, you walk right in, pull up a chair, and close the distance between you immediately.

These distances are maintained somewhat on the telephone. For example, if you call a customer or superior who is out of the office, you may hesitate to leave a message and ask them to return your call. Instead, you may offer to call back later. On the other hand, you would not hesitate to expect a colleague or subordinate to return your call as soon as the message was received.

☐ Voice Quality

The only nonverbal symbol that can be perceived over the telephone is voice quality. If your voice becomes high-pitched and loud, you show you are excited or anxious. If your voice is a low whisper, you show the conversation should be confidential. A low monotone tells your lack of interest. Excessive speed says you are under pressure or too hurried to be concerned about successful communication. Frequent hesitation reveals your lack of confidence or feeling of uncertainty.

■ METACOMMUNICATION

Metacommunication refers to the expression of ideas by implication. It is not so much what you say, but how you say it. Sarcasm and satire are forms of metacommunication. Many jokes rely on metacommunication for their humor. Here are some more subtle examples of metacommunication. Suppose someone says to you, "My, don't you look fine *today*." You smile and say "Thank you," but you walk away wondering "Does that mean I usually look pretty bad?" A supervisor might admonish an inexperienced clerk, "Please try to get *this* one right." Can there be any doubt about the implied "You have made at least one mistake in everything you have done so far." A manager may instruct a service representative, "Make a good impression on *this* customer," leaving no doubt that this one is more important than others.

You may have a problem with metacommunication when you are trying to be funny on the telephone. Metacommunication often requires nonverbal signals for the joke to get across—such as a certain gleam in your eye. Use extreme caution!

■ PERCEPTION

With our understanding of the symbols used in human communication, let us consider how these symbols are affected by perception. *Perception* refers to our point of view, which acts like a screen in both the encoding and decoding processes. This screening effect may cause us to interpret the same message or symbols in different ways, depending on who we are, how we feel, and what is going on around us.

Our interpretation of messages we receive is greatly affected by our self-concept, our experiences, and our roles in the immediate situations. One very good illustration of how the denotation of a single verbal symbol can differ in the perception screen of a sender and a receiver, each locked into a different role, can be found in this anecdote about Richard Nixon when he was President of the United States. The story is told by Dan Rather and Gary Paul Gates,[1] and the application is drawn from a textbook on organization behavior.[2]

> The President was working alone, very late at night in a hotel room while on a trip. He opened the door, beckoned to a waiting aide and ordered, "Get me

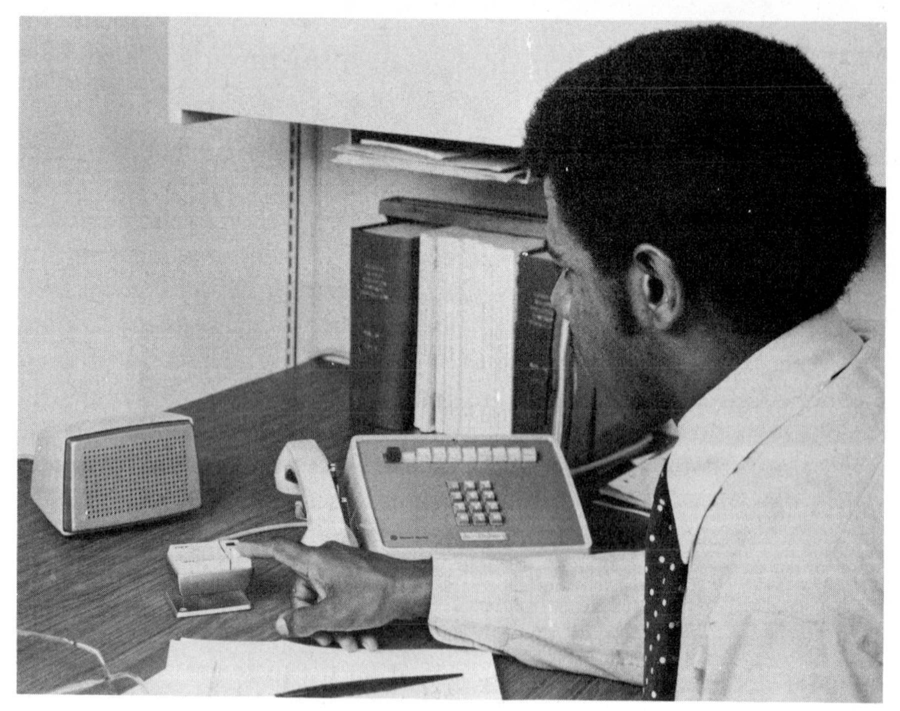

AT&T's 4A Speakerphone® used in conjunction with a 10-line key telephone. Photo reproduced with permission of AT&T.

[1]Dan Rather and Gary Paul Gates, *The Palace Guard* (New York: Harper & Row, 1974). Permission granted by William Morris Agency, Inc., on behalf of authors.

[2]Andrew D. Szilagyi, Jr. and Marc J. Wallace, Jr., *Organizational Behavior and Performance,* 2nd ed. (Glenview, Ill.: Scott, Foresman & Company, 1980).

coffee." The aide immediately responded to the request. Most of the activities of the hotel including the kitchen were not operating at such a late hour. Hotel personnel had to be called in and a fresh pot of coffee brewed. All of this took time and the President kept asking about coffee while waiting. Finally a tray was made up with a carafe of coffee, cream, sugar, and some sweet rolls and was rushed to the President's suite. It was only at this point that the aide learned that the President did not want coffee to drink, but rather wanted to talk to an assistant whose name was Coffee.

Perception is equally affected by motivation, needs, and feelings. A small request from a potential customer may be a welcome message for a commissioned salesperson, who sees every contact with a customer as an opportunity to make a sale. That same message would irritate a secretary who must interrupt progress on a rushed typing project to handle the phone call. The customer is an incentive to the motivation of the salesperson but an obstacle to the motivation of the secretary.

Finally, consider the effect of experiences and surroundings when a memorandum announces an impending merger. The merger may threaten an employee who lost a job in the merger of another company but be good news to the employee who owns stock in the new affiliate and expects its value to increase.

■ FEEDBACK

Now you are more aware of some reasons for problems in the process of human communication. With all these pitfalls, you may wonder how we manage to communicate successfully. The answer is feedback. If you review the illustrations, you will see a feedback line returning from the receiver to the sender. This feedback is the lifeline of successful communication. It is like the echo check or parity check in data communications.

Feedback can be as simple as a nod or as elaborate as a paraphrase of an entire message. The responses in the illustrations show how feedback uncovered discrepancies in the encoding and decoding process. The best way to use feedback as a means for communicating successfully is to keep talking *and listening* until you are sure you have achieved total understanding. In telephone conversations, you must rely totally on verbal symbols for feedback. Blank looks and quizzical expressions simply cannot be seen. Thus listening becomes even more important.

LISTENING

Too often we forget that talking is only half of communication. Listening is the other half, and it may well be the most underestimated skill in all human relations.

Pause here for just a moment and think of several people who are good listeners. Jot their names in the margin. Chances are you listed several people about whom you feel very good. We tend to like people who listen, because they enhance our self-esteem, make us feel our ideas are important, and help us satisfy our basic need to communicate.

Fortunately, listening is a relatively easy skill to develop. First, it is important to realize that all of us can think much faster than we can speak. Our frequently tangled tongues attest to that. As listeners, then, we have some extra time. What we do with this extra time determines our success as listeners. Here are some things to remember.

1. *Be an active Listener*. Give a lot of feedback! Check out the accuracy of the symbols that pass through your perception screen. Be sure you are decoding correctly. Make comments to show your understanding (or to reveal your misunderstanding). Don't be afraid to ask questions.
2. *Be a thinking listener*. Analyze the speaker's perception screen. Try to understand his or her motivation and point of view, even if you do not agree. Evaluate the speaker's reasoning and review the facts you have heard.
3. *Be a receptive listener.* Keep an open mind. Don't clog your perception screen by jumping to conclusions. If you start to plan your response while you are listening, you will probably miss some key information.
4. *Be an alert listener*. Keep an alert posture and hold your attention on the speaker. Show him or her that you are interested. Even if you are talking on the telephone, smile and nod right along with your affirmative responses to what you hear. If you are talking face-to-face, maintain eye contact.
5. *Push distractions out of the way*. Even taking notes can be distracting. Reserve note-taking for facts and figures; do not attempt to record concepts and ideas.
6. *Pay compliments*. Demonstrate appreciation for both the message and the speaker wherever possible.

SOME DELICATE SITUATIONS

Jokers and Windy Talkers

The comedian may either brighten your day or annoy you because you are under pressure of a heavy work load (or because the joke is not funny). The incessant talker will nearly always annoy you. However, both the joker and the windy talker deserve your courtesy. You must be able to gain control of the call after either laughing politely or listening to at least a few sentences.

The best way to gain control of the conversation is to watch for an opening and enter where you can *agree*:

> Yes, they certainly should do something about inflation. I can help you with your present problem, however. What is your question about your bill?

or

> Yes, we've had enough rain this spring. For the trouble you're having with your convertible top, though, let me connect you with someone in our Automotive Department.

or

> It certainly is difficult to raise children these days. Right now, however, how can we help you with the tricycle?

It is not dangerous to ask another question if the question helps the caller to focus the conversation directly on the purpose of the call.

☐ Irate Callers

The irate caller is not quite the same as the windy talker. This person needs to vent anger before service can occur and the good will of the company can be restored. Try to let him or her talk, and listen sympathetically until you hear a lull in the storm. Then, get the call focused on a problem that can be solved, and tell the caller exactly how you can help.

☐ Reaching an Impasse

Now, let's be realistic. the customer is not always right. Once in a very long while, it becomes necessary to terminate a call even though the caller insists on continuing. If you are absolutely sure you have found yourself in such a situation, you can say the following:

> I'm sorry, but we can't seem to get together, and there is no point in our continuing. Excuse me please; I must leave the line. Good-bye.

Then place the receiver on the switchhook *very gently*, take a deep breath, and continue with your job duties.

☐ People in Panic

The caller who is in a panic for any reason needs to connect with someone who has a cool head. If you are handling such a call, you should first assure the caller that he or she has done the right thing to call and then give firm instructions to the caller to remain calm. The calmness in your voice will be contagious. Help the caller focus on an accurate description of the problem and a statement of the essential facts. Then explain exactly what you are going to do to help and remain on the line giving reassurance until help arrives.

☐ Giving Bad News

Bad news that is really serious is frequently handled by letter or in person by a trained trouble shooter. However, you may occasionally be

Attendant's console for Dimension® PBX system. Photo reproduced with permission of AT&T.

the bearer of disappointing information. When this happens, use good human relations, but make the message absolutely clear.

One good strategy is to try to state the bad news positively:

> Mr. Sims will be out of the state on the day of your meeting, and he asked me to tell you how disappointed he is that he will miss the opportunity to speak to your members.

A second strategy is to surround the bad news with good news:

> We have your order for telephone timers ready to ship. However, since it would take several weeks to establish your credit, we would like to have your permission to ship them COD. If that is all right, they will be on your shelves by Monday.

☐ Bombardment

You are experiencing a minor case of bombardment when two lines ring at the same time. It is easy to handle, however. You simply answer one line with the proper greeting and identification and say, "Please hold." Allow the caller to respond before establishing the hold. Then answer the second line. If it is a call you can handle quickly, allow the first call to wait until the second call is complete. If the second call will be lengthy, explain that you have a call waiting and give the caller the option of either holding or leaving a number where you can call back.

If you are walking out of your office to get some information for a caller you have on hold and you spot someone headed toward you whom you know is waiting for data from you, hear your phone ring again behind you, and encounter your boss wearing a purposeful frown, you are experiencing an acute case of bombardment. It happens to every knowledge worker at least once a week. Do not panic, and do not allow yourself to appear victimized. Keep you head, be prepared to explain your situtation objectively, and take the following steps:

1. Rely on your colleague to recognize what is happening to you and ignore him or her.
2. Explain to your boss that you have a call on hold and another line is ringing. (The telephone takes precedence even over bosses.)
3. Answer the ringing line, asking the caller to hold.
4. Check back with the caller already on hold.
5. Laugh.
6. Make a joint decision with your boss about whether you should ask the callers to wait or make a note to see your boss as soon as the calls have been handled.

Yes, you should laugh. Business, you will find, is populated by people. People are just like you. Some days you muddle through a mass of miscommunication, while other days go smoothly and leave you feeling quite accomplished. Both kinds of days will be full of telephone conversations with people.

QUESTIONS FOR REINFORCEMENT AND DISCUSSION

1. Describe the model of human communication and compare it to the model of data communications in Chapter 5.
2. Name and define the three types of symbols in human communication.
3. What are some of the factors that cause the sender and receiver to perceive the same symbol differently?
4. What is the role of feedback in the communication process?
5. Why is listening an important technique in good human relations?
6. Name five ways in which you can be a good listener.
7. Describe a situation in which you miscommunicated with someone because something went wrong in your perception screen.
8. What is the difference in technique between a call involving a windy talker and a call involving an irate caller?
9. What do you do first when you receive a call from a person in panic?
10. Consider some of the differences in the perception screens of the following:
    a. A beautiful, young girl communicating with an ugly, old man.
    b. An executive vice president communicating with an office page.
    c. A commissioned salesman communicating with a commissioned saleswoman.
    d. A customer placing a $1 million order with a receptionist making $5 per hour.
    e. An Air Force colonel communicating with an airplane mechanic.
    f. A computer programmer communicating with a middle-aged housewife.
    g. A research scientist communicating with a student writing a paper.

A CASE OF CONFUSING SYMBOLS

Identify the communication problems in the following call:

Operator: Good morning. Barker's Department Store.

Customer: I want to buy some files.

Operator: Thank you, I'll connect you with Hardware.

Hardware: Hello, this is the Hardware Department.

Customer: Hello. I need some files.

Hardware: Metal or wood?

Customer: Well, I guess metal would be all right.

Hardware: What size?

Customer: I don't know. They're just little boxes.

Hardware: Boxes! Oh, you don't want tools. You want files.

Customer: Yes. Files.

Hardware: Just a minute. I'll switch you to Stationery Supplies.

Stationery: Hello, this is the Stationery Department.

Customer: Hello, I need some files.

Stationery: Legal or letter size?

Customer: I don't know; they're just little boxes.

Stationery: Oh. You don't want folders; you want file boxes.

Customer: Yes, file boxes.

Stationery: What are you keeping in them?

Customer: My cassette tapes.

Stationery: I see. Well, we don't carry those files here, but you might find them in the Appliance Department. They have stereo equipment there. Just a moment and I'll transfer you.

Appliances: Hello, this is the Appliance Department.

Customer: I need some file boxes for my stereo tape cassettes.

Appliances: I know exactly what you mean. Just a moment and I'll get our Record and Tape Department clerk over here to help you.

Clerk: Hello. I understand you want some of our File-rite Cassette Cases. Did you want the 12-cassette size or the 24-cassette size?

Customer: I didn't know there were two sizes.

Clerk: The 12-cassette size is $5.99 and the 24-cassette size is $9.99. How many cassettes are you filing?

Customer: A lot. I'll take two of the 24-cassette size.

Clerk: All right. Will this be on your charge?

ROLE-PLAY REPLAY

Choose a partner and replay one of the entire conversations from the role-playing exercises in Chapter 6. This time, record your conversation on a tape cassette. What you will hear is essentially how you sound on the telephone. Analyze your telephone communication ability by answering these questions.

1. Does my voice have a pleasing tone, not harsh, nasal, or too loud, but loud enough to be heard easily?
2. Do I speak clearly and distinctly and at a speed that makes me easy to understand?
3. Do I speak with an accent that I should minimize?

4. Do I express my thoughts concisely, with well-constructed sentences and without a lot of slang, "you know's", or "uh-huh's" that sound repetitive and tedious?
5. Do I sound like I am interested in the conversation and in the person with whom I am speaking?

A CASE OF COMMUNICATIONS EMBARRASSMENT

Roslyn Green is responsible for the entire sales operation for the Appliance Division of Sims Products, and she has a serious communications problem. She talks on the telephone to so many customers and associates every day that she can't keep them all straight. The salespeople on her staff handle the actual sales, but she becomes involved with nearly all accounts at one time or another because she is responsible both for generating new business and for public relations.

In addition, she is extremely active in civic affairs. She is on several political action committees, is an officer in her professional women's association, and is a den mother for her son's Cub Scout Troop.

She finds she frequently receives calls from people who expect her to remember who they are, what they want, what she has discussed with them before (sometimes many months ago), and what they should do next.. Everyone always seems to remember her well, but she is too often embarrassed because she either can't remember someone at all or she confuses one person with another. She recently thought a customer was a Cub Scout when she answered a telephone message! She is suffering from a severe case of *information overload.*

She feels that she could handle everything more effectively if she could refer to some sort of chronological record of her telephone conversations, which would help her recall them; she would also like a way to make notes that remind her to follow up certain conversations with specific action. She has asked you to design a communications log that will accomplish this objective.

The log you design should be styled like a calendar, perhaps similar to an appointment book. You might allow a full page or a half page for each day of the week. Each day should have a space for reminders of conversations to be followed up that day and a space for recording conversations that occurred that day. Information entered for each call might include the name of a person, the name of a company, a complete telephone number, and subjects discussed or conclusions reached. Put yourself in Ms. Green's shoes and imagine yourself making entries in the log and referring to it each day to help you talk on the telephone with confidence and competence.

8 Communication Strategies for Business

Many of your telephone conversations accomplish essentially the same task, such as requesting information, taking orders, or setting appointments, over and over again. This chapter contains some useful strategies for situations that occur frequently.

▪ STARTING

You should always begin by identifying yourself completely. Any time you assume you are easily identified, you are taking a risk. If you are starting a conversation with someone you call frequently, you might omit your company affiliation. If you are talking with someone you know quite well or someone who is likely to recognize your voice, you might omit your last name. If you are calling someone you do not know or someone who may not recognize you by name or company affiliation, identify yourself slowly and distinctly and then recall an event or a mutual acquaintance that will identify you:

> **Hello, Mr. Sims. My name is John Hendrix. I represent Precision Drilling Products. I met you at an SEG meeting last month in New York.**

or

Hello, Mr. Jones. I'm Susan Clay with the City Attorney's office. Howard Marks suggested I call you because he thought you could help me locate some information I need.

When you are sure you have properly identified yourself, express interest in the person you called.

How are you today?

or

How was your holiday?

or

I know you're very busy, and I'll take only a moment.

or

I read in the *Journal* about your new product line. How is it going?

Don't overdo this step with flowery compliments. Just the expression of interest is complimentary enough. Give the person you have called a pleasant interlude in which to become acclimated and receptive to your call.

Now state your purpose, clearly, concisely, and directly. You may even wish to include a complimentary statement of why you chose to call this person rather than someone else.

We are looking for a reliable company to repair the exhaust system for our smelting furnace, and you have been recommended.

or

Our school's business club is planning a ski trip during spring vacation, and we think your lodge would be a great place for it.

or

I'm calling about the committee being formed for consumer action in the investment-fraud activities you heard about on the news this week. I know you were recently involved in a similar case that was concluded quite successfully.

■ ASKING

Asking for information or assistance should be easy. However, if you are not prepared in advance, you can get incomplete information or information you do not want, or even turn a simple task into a nightmare. Start by understanding your need. Formulate it carefully in your mind. If you need an item or product, be as specific as you can about its identification. Try to find out the model number, size, style, color, and price. Carefully estimate the quantity you will need, how long a certain supply will last, how you will store the quantity purchased, and how it will be used. You can refer to a former order or a specific advertisement where you learned about the product for many of these facts.

We need four dozen of your number 9 magnetic disks. We want the 5-inch size for our personal computer. We saw them advertised in this month's *Computer News* magazine for half price. We expect to use about a dozen each month.

If you need information or service, be prepared to explain the reason for your inquiry.

I am arranging a breakfast meeting for our 15 sales representatives on Friday, June 2. Most of the representatives live on the south side of town. Can you recommend a hotel or restaurant that can give us a private room as early as 7:30?

Before you pick up the telephone, be sure you can answer all six questions of fact: who, how, where, what, when, and why. Then picture yourself handling your own call, ask yourself what must be done, and be sure you are ready with all the information required.

Try to understand how your request will affect the person you are calling. The following are some of the concerns that run through a person's mind when on the receiving end of a request:

- Why am I the one getting this request?
- Why does this person want to know this information?
- Should someone else handle it?
- Will anyone object if I comply with this request?

Feature package 5 for Dimension® PBX system. Photo reproduced with permission of AT&T.

- ○ What will it cost me to do this or give this?
- ○ Will I have time to do this?
- ○ How often am I going to get this kind of request?

If you anticipate any concerns that may arise, you can be prepared to show that you understand the other person's situation. Then make it as easy as possible for the person to provide the information or service you need.

> All our members are involved in the medical profession, so they will understand the technical nature of your presentation.

or

> If you can just give us a basic outline, we can fill in the details.

or

> If I am not here when you call, you can just leave the information with my secretary.

or

> My client understands how much is involved in this sort of project, so there is no great time pressure.

■ TELLING

People on the receiving end of an information exchange have concerns similar to those on the receiving end of a request.

- ○ Why am I getting this information?
- ○ How am I expected to react to it?
- ○ What am I expected to do with it?
- ○ How long will I need to remember it?
- ○ Will anyone object to the fact that I know this?
- ○ Should I tell someone else about it?
- ○ How much of this should I write down for future reference?

Your opening identification, expression of interest, and statement of purpose should be carefully planned to allow time for the recipient to get ready to write.

> This is Lucy from the insurance office. I have the information you requested about your claim. Do you have a pencil handy?

Be sure you present the information in an organized, logical manner. If the information is complicated, plan ahead and number each item. You may even want to make notes so you do not forget anything. Take it slowly, especially when giving numbers, names, dates, and amounts of money. Listen for feedback. Be sure you are understood. Allow plenty of time for response from the person you call. Use jargon only if you are sure it will be understood. Do not assume that the person requesting information knows as much as you do, and allow a reasonable amount of time for it to sink in. Do not condescend, and do not give the impression that you are superior because of your knowledge. Finally, close with an offer of future assistance if it is required.

■ PERSUADING

You are persuading when you bring someone around to your point of view. You may ask that person to do something he or she would not otherwise be required to do, to give you some information not readily accessible or of a delicate nature, or to adopt your way of thinking. The key to persuasion is "you attitude." Take the time to understand the other person's position, needs, and motivation; then explain your purpose in terms of the other person's interests. Be sure the desired results are not threatening to the other person, and emphasize how they are beneficial.

> Our CPA firm is conducting an audit for the Bothwell Department Store. We just need you to verify that you have a credit account with them and that the balance is correct. (Can you see how this approach avoids threatening a customer who is about to be questioned about his or her debts?)

or

> We are conducting a survey of customer reaction to our services so that we can serve you better. Would you mind answering 10 short questions? (This request relieves the person of the feeling that he or she may not have time for such a questionnaire.)

For a more involved request, you may follow the entire persuasive process. Consider this example of a telephone solicitation. Notice that Amy starts out by getting positive attention to her purpose.

> Hello, this is Amy Johnson. I am a member of the band at Central High School. Did you see in the newspaper that we have been invited to appear at the Independence Day Parade at the State Capitol this year?

Then she describes her purpose with appeal to the interests of the person she is persuading before she actually states her request.

> We are really excited about it and we want all the local businesses to have a chance to share in our success. We are conducting a newspaper drive to raise funds for new uniforms, and we hope Sims Products will participate.

Now that she has created a personal interest for the individual she is calling, she can state her purpose directly.

> Would you be willing to have your employees save all their old newspapers and bring them to your location for us?

Then she makes it easy for the company to participate.

> One of the band members will pick up the papers you save every Friday morning for the next 6 weeks.

Then she closes with a summary of what she will do and an expression of appreciation.

> I will put your name and address on the pickup list for our newspaper drive, and one of our members will stop this Friday morning. All the participants will be announced in our school neswpaper, too. We are glad you could help.

A need for interdepartmental cooperation might include all the described ingredients in a brief conversation. Persuading the department manager to help could be done like this.

> Next week, we are switching over to a new telephone system that will add some important conveniences to your phone service. We need to have your people ready for installers to come on Friday evening and work over the weekend. Could you ask everybody to clear their desks and check for easy access to all existing cables and outlets? We want to be sure none of your working areas are disturbed.

■ NEGOTIATING

Negotiating occurs when two parties are trying to agree on something. It is important to recognize that the two parties are seeking resolution, not trying to be winners of an argument. There is clearly a conflict, but both have something they want to gain and something they are willing to concede. The key to successful negotiation is an understanding of the positions, motivations, and needs of both parties. The final solution will satisfy some of the needs of both and include some concessions from both. To be successful at negotiations, you must be willing to keep talking, keep listening, and keep exploring possibilities for trade-offs and solutions.

■ TIMING

In addition to the common-sense tips about when to place calls across time zones, some guidelines can be established for calling people in certain industries, professions, or jobs. You will discover these times for your particular situation through experience. However, Mutual of New York did some studies and came up with the following suggestions, primarily for insurance prospects:[1]

Prospects	*Best Time to Call*
Chemists and engineers	Between 4 P.M. and 5 P.M.
Clergymen	Thursday or Friday
Contractors and builders	Before 9 A M. or after 5 P.M.
Dentists	Before 9:30 A.M.
Druggists and grocers	Between 1 P.M. and 3 P.M.
Executives and business heads	After 10:30 A.M.
Housewives	Between 10 A.M. and 11 A.M.
Lawyers	Between 11 A.M. and 2 P.M.
Merchants, store heads, and department heads	After 10:30 A.M.
Physicians and surgeons	Between 9 A.M. and 11 A.M.; after 4 P.M.
Professors and schoolteachers	At home, between 6 P.M. and 7 P.M.
Public accountants	Anytime during the day, but avoid Jan. 15 through April 15.
Publishers and printers	After 3 P.M.
Small-salaried people and government employees	Call at home
Stockbrokers and bankers	Before 10 A M. or after 3 P.M.
Prospects at home	Monday nights between 7 P.M. and 9 P.M.

■ KEEPING AN APPOINTMENT CALENDAR

The appointment calendar, especially in the office of a professional person such as a doctor, attorney, employment counselor, or salesperson, is a vital tool of the business. Not only does it guide the activities of each day, but it also becomes a chronological record that may be used for billing, statistical analyses, and management planning.

If you are in charge of the appointment calendar, your two most important responsibilities are to (1) make the calendar the most accurate and effective tool possible, and (2) treat each and every caller with the utmost courtesy. In order to serve the person who calls for an appointment, you must first understand the need for the appointment

[1] Larry A. Arredondo, *Telecommunications Management for Business and Government* (New York: The Telecom Library Inc., 1980), p. 225.

and then get an idea of about when the caller desires the appointment. You can use the probe questions suggested for the full screen to learn the caller's need, and then try to give a choice when you actually set the time and date.

> Would morning or afternoon be better for you?

> The earliest opening I have on Thursday is 11 A.M. If you prefer an earlier time, I can give you 9 A.M. on Monday.

Next you must assure a complete and accurate exchange of the following information.

The caller must know:

- Name of the person to be seen.
- Address of the location, including adequate directions, information about public transportation, and parking instructions.
- Date and time of the appointment.
- Any special instructions that may be required, such as what to do ahead of time or what to bring.

Your records must show:

- Name and address of the person coming in.
- Name of the person to be seen.
- Date and time of the appointment.
- Reason for the appointment.
- Identification of the caller, such as file number, account number, or case number.
- Telephone number of the caller in case a change is necessary.

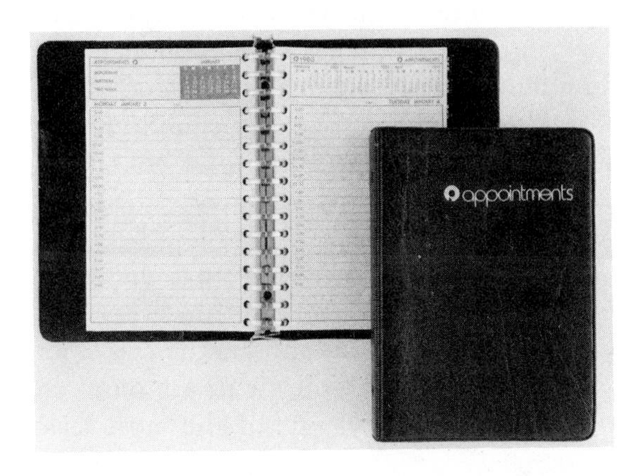

Photo of spiral-bound appointment calendar provided through the courtesy of Rediform Office Products.

■ MAKING AND TAKING ORDERS

A telephone order-taking procedure is generally guided by a form, which is filled out by the person taking the order. As long as the form is filled out completely and accurately, the order can be filled promptly and correctly. However, the communication that occurs between the customer and the person taking the order should not be allowed to become mechanical. Each caller must be treated with all the warmth and courtesy that would be extended to a customer who comes into a store or showroom.

If you are taking an order, remember that customers do not have copies of the form. Do not expect the customer to have all the information you require. Be prepared to search your files for account numbers or stock numbers and ask for information the customer may not realize is important.

> Yes, Mr. Jones, you need six dozen men's dress shirts, two dozen size small, two dozen size medium, and two dozen size large. Those are stock number 548. There is an extra-large size too. Do you want to stock those also?
>
> Did you want short sleeves or long sleeves?
>
> The long sleeves come in two styles, the regular cuff and the French cuff, which is becoming very popular these days. Do your customers have a preference, or would you like to stock both?
>
> The price on these shirts in the quantity you have ordered is $6.15 after your regular discount, if you'd like to make a note of it.

Some of the information usually required for telephone orders includes complete identification of the company, with shipping and billing addresses and an account number, a purchase-order number, and current date, followed by the quantity desired and a description of the item (including details such as stock number, size, style, and brand). Price information may include the unit price and the total price, plus information regarding taxes or discounts that apply. Finally, the order may specify the method and date of delivery.

Be sure you leave the customer feeling confident that the order will be filled according to expectation. Repeat and confirm every fact and detail. If possible, check stock availability and inform the customer of any possible delays. Ask the customer when the items are required if that information is not volunteered. Finally, one of the most reassuring steps you can take is to give the customer your name, so that you can be contacted if any questions or problems arise. Your name or initials will, of course, appear on the order form.

■ HANDLING COMPLAINTS

In spite of the best efforts of millions of competent people employed in thousands of reputable firms, things still go wrong. If there is a problem with goods or services provided by your company, you may be asked to handle a complaint. In business, this is not necessarily an unpleasant task, because most people realize that shouting and anger do not get the job done. Moreover, most companies make fair, if not liberal, adjustments in the interest of maintaining good will. An adjustment may be a cash refund, a credit refund, an exchange, or a replacement.

If you handle complaints, it will be your job to know and understand how to apply company policies regarding adjustments and to treat every caller with utmost courtesy, even under stress. Here are the basic steps.

1. Listen patiently and sympathetically to the caller's description of the problem and statement of the desired adjustment. Your ability to communicate may be put to an extreme test when you try to get a clear picture of the problem out of a muddled account of what has happened! If possible, try to identify the transaction involved by date and order number.
2. Try to agree with the customer whenever possible. By doing so, you make the individual feel you are on his or her side and weaken defenses that may be obstacles to the task of mutually solving the problem. Also, avoid placing the blame. The solution is more important than the fault. Moreover, you do not build the customer's confidence in your firm when you are critical of your colleagues.
3. Express concern for the customer's situation and confidence in your company's ability to stand behind its product or service. A brief apology here also goes a long way toward building good will.
4. Explain the company policy and how it will be applied in this case. Be sure the customer knows exactly when the adjustment can be expected. Above all, don't make promises you or other employees cannot keep.
5. Follow through after a reasonable time to be sure the commitment you made is carried out and the customer is satisfied. This will really cement the good will you built when you handled the complaint well.

Here is an example of an adjustment handled well by both parties. Notice how a situation that could have resulted in accusations and aggravation is turned into a solution for the buyer and additional business for the seller.

Seller: Good morning, Tri-state Business Products. This is Susan speaking.

Buyer: Hello, this is Bill Garcia with City Business Supplies. We have a problem with the calculator order we placed last month.

Seller: I'm glad you brought it to our attention, Bill. What seems to be wrong?

Buyer: Well, your sales rep told us that the calculators came without batteries, so we bought a gross of AA's from a supplier, who won't take them back. Then when we got the calculators, they all had batteries already, and now we're stuck with 144 batteries going dead on our shelves.

Seller: I'm sorry you're having this hassle. Will you need them later to replace the originals?

Buyer: Not really. Our customers generally buy their replacements somewhere else, and we don't usually carry batteries.

Seller: I see. Well, in that case, I think the best way for us to help is to give you credit on the return of the batteries to us. When do you expect to order again?

Buyer: Probably in the next week or two.

Seller: If you would like, I can go ahead and place the order now. I can process the credit at the same time.

Buyer: OK. Put us down for 125.

Seller: Fine. Let me see, now, your file says you use the model 316 with the display. Is that correct?

Buyer: Yes, that's the one.

Seller: Let me call you back after I figure the amount of your credit and the total cost of the calculators. You can give me your purchase-order number then.

Buyer: Fine, I'll wait to hear from you. Good-bye.

Seller: Good-bye, and thanks again.

■ SELLING ACTIVITIES

The telephone plays a major role in three different selling activities: inquiry calls, prospecting, and telephone sales.

□ Inquiry Calls

Probably the most important telephone call is the buying inquiry. This caller is asking for information about the company's product or service and indicating an interest in buying. When you answer such a call, you are hearing the voice of opportunity. Many companies refer to them as *leads*. Much of the organization and effort of many employees in the firm is devoted to turning leads into profitable sales.

If the company has a formal advertising campaign, the advertising purchased in various media, such as newspapers, magazines, radio, and television, costs a great deal of money. A well-managed company analyzes such costs to determine how much money is actually spent to get each inquiry. (Total advertising dollars per month divided by the number of leads each month gives the cost of each lead.) Companies have found that these leads can cost over $100 each. Careful records are kept of all calls received, where they are routed, and if they result in sales.

If your duties include handling such calls, you will be given thorough instructions concerning what you should tell the prospective customer, what you should ask the prospective customer, and what records you are responsible for keeping. When you stop to consider the costs of getting such calls and the subsequent costs of converting them into sales, you will always be on your toes with your best service attitude when you answer the phone.

□ Prospecting

No doubt you have always thought of prospecting as looking for gold—and you are absolutely correct. In sales activities, gold is the potential buyer. Successful salespeople look for prospects everywhere. Many sales operations find telephone prospecting just as effective as advertising in locating potential buyers. With costs of paper, printing, and postage going up, the telephone competes favorably with direct-mail advertising. Moreover, telephone prospecting is sometimes more effective than media advertising because it can be directed to specific segments of the buying public rather than to the general public. Finally, the telephone can be more effective than either media advertising or direct mail because of its personal impact.

In a telephone prospecting operation, employees do nothing but call lists of people or companies to find possible buyers. WATS lines and switched-voice services are used to cover wide territories. The calls may be made by using the regular telephone directory or criss-cross directory that is organized by street address to permit canvassing of specific neighborhoods, or they may be made by simply dialing every number of an exchange in sequence. Other prospecting operations are more selective. Calls may be made from specific sections of the yellow pages to reach certain industries or retail establishments, from trade directories, or from prospect lists, which may be purchased or rented from mail-order houses or subscription services. Many companies find lists of inactive customers to be an excellent source for prospecting.

Once a prospect is contacted, the next steps are to qualify the buyer and set an appointment in which a sale can be made. *Qualifying the buyer* means determining if the person has a valid need for the product or service, has the authority to make the purchase (is the head of a household or an authorized purchasing agent for the firm), and has the ability to pay. This step requires delicate probing. Finally, setting the appointment requires a little persuasion, plus the skills suggested for maintaining the appointment calendar.

☐ Telephone Sales

Sometimes, a telephone salesperson goes beyond finding leads, to actually make the sale and turn in an order. Selling over the telephone has advantages over selling by appointment for both the buyer and the seller. It saves the buyer's time and it reduces the seller's travel costs. Telephone selling, however, involves the same techniques as personal selling. In order to be effective, the salesperson must have an in-depth knowledge of the product and be able to assess the needs of the buyer in terms of that product. A professional telephone salesperson is carefully trained and is often given a written sales "pitch" to follow.

■ COLLECTING MONEY FROM ACCOUNT CUSTOMERS

If you are to be successful with the telephone as a collection tool, you must understand the nature and purpose of credit. The objective of credit is to increase sales. Business managers realize that more sales can be made if they are willing to wait for payment and assume some risk. Some credit expenses and losses are inevitable, but as long as those expenses do not exceed the extra profit generated by the increased sales, credit is good business practice.

The word credit comes from the Latin root *credere*, which means *to believe*. Applied to credit activities, it means "I believe you will pay," and it must be the underlying theme of your communication with account customers. Your objectives are to collect money due while maintaining the good will of the customer and keeping the door open to future business. These goals may seem to conflict, but they are compatible if you understand and maintain the proper attitude toward the debtor.

To understand the debtor, you must first realize that a customer who buys on credit is aware of debts and expects to be asked to pay. Do not be timid or apologetic about calling. Get as much background information on the account as you can. Has payment been prompt in the past? What attempts at collection have already been made? What is the name of the person who is actually responsible for making the payment? What are your company's credit policies for this type of account? Finally, you will need the details of the account, such as purchase-order numbers, dates of purchase, amounts, and authorized purchasing agents.

Start the call properly by identifying yourself, expressing an interest in the person you reach, and then stating your purpose clearly and directly:

> I am calling about your account with us.
>
> or
>
> I am calling about your past-due balance.

Now, be prepared to listen to the customer's response. Then based on that response, either make a direct request for payment or suggest a payment plan.

> When can we expect to receive your check?
>
> or
>
> Can you make a partial payment now and then send the remaining balance within 30 days?

Wherever possible, request payments in terms of the customer's needs, interests, and pride.

> You have always paid on time in the past. How soon do you expect to be able to clear your present balance?

or

> We want to continue selling to you as an account customer. Can you make monthly payments and clear the balance within 90 days?

Finally, keep careful records of all your conversations so that you can follow up after a reasonable amount of time has elapsed.

■ CONCLUSION

In all your business communications, whether you are buying or selling, asking or receiving, or basking in success or tackling a tough problem, remember that you are always talking to people. People respond quite favorably when they are treated with courtesy and respect. Yet there are days when pressures and problems make it difficult to smile, and everyone's patience runs thin. You are sure to find, however, that even when the going gets really rough, people communicating with people can always find a way to success.

QUESTIONS FOR REINFORCEMENT AND DISCUSSION

1. Describe a good way to start a telephone conversation with someone you have never met.
2. What are some things you should consider before placing a call in which you are going to make a request? In which you are going to give information?
3. What are some of the questions that run through a person's mind when receiving a request? When receiving information?
4. How are negotiating and persuading different from each other?
5. How can you make handling complaints a pleasant task?
6. Describe a telephone call you made in which you complained about a product or service and asked for an adjustment. Do you feel your call was handled properly?
7. What are the two functions of an appointment calendar?
8. Distinguish among the three different selling activities in which the telephone plays an important role.
9. Describe a call you received from someone trying to sell you something over the telephone. How did you feel about receiving the call?
10. When making a collection call, how should you view the debtor?

TELEPHONE TASKS

1. Your first task is to call the 16 regional directors of your company and inform them that they are to participate in a conference call with the vice president on Tuesday of next week. Plan what you will say to the directors and try to anticipate any questions you might be asked. The vice president wants the call to

APPOINTMENTS FOR July 16 19 XX

TIME		NAMES	PHONE	COUNSELOR	SHOWED	AMT
8 A.M.	00					
	15					
	30					
	45					
9 A.M.	00					
	15					
	30					
	45					
10 A.M.	00					
	15					
	30	Shari Washington	no phone	Mr. Bennett		
	45					
11 A.M.	00					
	15					
	30					
	45					
12 P.M.	00					
	15					
	30					
	45					
1 P.M.	00					
	15					
	30					
	45					
2 P.M.	00	Lauren Straus	555-1279	Mr. Bennett		
	15					
	30	Robert Smith	555-8410	Mrs. Arnold		
	45					
3 P.M.	00					
	15					
	30					
	45					
4 P.M.	00					
	15					
	30	Arlene Corders	555-1984	Mrs. Arnold		
	45					
5 P.M.	00					
	15					
	30					
	45					

SALESMAN	BARKER'S DEPARTMENT STORE TELEPHONE ORDER FORM DATE ______ 19__	OUR ORDER #
TERMS	SOLD TO ______	CUSTOMER PO #
DELIVER BY (DATE)	ADDRESS ______ ORDERED BY ______ PHONE ______	DELIVERY METHOD

STOCK #	DEPT #	QUANTITY	DESCRIPTION	UNIT	PRICE	DISCOUNTS	EXTENSION

originate from the conference room in your New York headquarters so that several other vice presidents may be included. The teleconference will begin promptly at 4:30 Eastern Standard Time. Regional directors are located throughout the United States and Canada. The subject of the call will be sales forecasts for the last quarter of this year, and each director is expected to have his or her forecasts finalized for the meeting.

2. You have volunteered to buy a birthday gift for the boss this year. The staff has indicated that a leather briefcase would be a good idea. There are 15 people on the staff, and each person will contribute about $10. The birthday is only 2 days away. You decide to call some stationery and department stores to find the best buy for the money. Make a list of the questions you will ask.
3. Assume you are in charge of the appointment calendar for someone at your office or school. Your teacher will play the role of a person calling for an appointment. Role-play the call from start to finish, filling out the appointment form given on page 161. Be sure you exchange all the information the caller must know and your record must show.
4. You have been elected treasurer for your professional organization. Therefore, it is your duty to call all the members who have not yet paid their annual dues and collect. Dues must be paid before the 15th of this month for members to be included in the association's directory. How would you do each of the following?
   a. Identify yourself.
   b. Start the conversation.
   c. State the purpose of the call.
   d. Make the directory listing an incentive.
   e. Ask for the money.
   f. Respond to the answer if it is yes; if it is no.
   g. Close the call.
5. Assume you are the salesperson at Barker's Department store and continue the conversation from Chapter 6 in which a customer wants to order two 24-cassette cases and charge them to his or her account. Your teacher will play the part of the customer. You are to complete the order form given on page 162.

A CASE OF COMMUNICATION OVERLOAD

This morning, Mr. Sims received a phone call from a long-time friend, one of Sims Products' first customers, a customer who has remained loyal through the years and who continues to order a fairly large volume of auto parts. The customer called to complain about the order takers in the Automotive Sales Department.

"It just seems to happen every time I call," he reported. "When the phone is answered, I can hardly tell what company I've reached because the receptionist says the name so fast. Then, I get this long silence on hold, and I'm not sure whether my call is going anywhere or not. By the time the order clerk gets on the phone, we're both pretty frayed. It looks to me like you just don't have enough people to take all the calls."

The customer went on to say that the whole order process seemed like a tedious burden to the order clerk. The questions were asked mechanically, as if they were coming from a machine gun. It is as though the customer were a real imposition. The only hopeful report the customer could make was that the orders were still filled with the usual Sims efficiency; deliveries were made within 2 days. "But just the thought of calling to place that order is a real drudge," concluded Mr. Sims' friend.

Mr. Sims is extremely concerned because he knows that those order clerks are the most important public relations contacts for the successful repeat business the company has enjoyed for many years. He asks Mr. Carson to investigate and make recommendations. What do you think Mr. Carson should do? What do you think Mr. Carson will recommend?

9

Costs and Controls

"You're not talking too much; you're just paying too much!" is from an MCI television commercial.

We call the telephone the most familiar tool of management. We say that no other machine can be found as universally as the telephone in business and government. Finally, much of this book is a discussion of how the telephone helps people to communicate and how communication helps people to get the job done. Therefore, it follows that a discussion of costs and controls should not be devoted to how to use the telephone less but rather to how to optimize telephone usage and expenditures. We are optimizing telephone usage when we have the best possible communications support for a reasonable cost—a goal that is easy to state but difficult to attain. Just exactly what is the best possible communications support? Just exactly how much is a reasonable cost? These questions are especially difficult to answer in today's rapidly changing environment.

Management attempts to answer these questions and achieve the goal of optimum communications services by studying the sources of telephone costs and the charges for telephone services. More sophisticated techniques for reporting and analyzing usage and charges are available today than ever before. The results of these studies are then the basis for decisions regarding equipment, services, and procedures.

■ SOURCES OF TELEPHONE COSTS

Sources of telephone costs may be studied effectively in three different categories: use, misuse, and abuse. We encourage the use of the telephone and study it carefully to be sure adequate services are provided. Telephone usage improves employee morale and productivity, as well as customer service.

Misuse is another story. It usually occurs because people do not have adequate knowledge of the equipment or telephone system of the company. For example, an employee may make a toll call on the wrong FX line, place an unsuccessful and wasteful long-distance call in a distant time zone when the windows are closed, or fail to notify the operator when a toll call is misdialed and the company is entitled to credit for the call. Most misuse can be eliminated if employees are well informed and trained in the use of the company's communications services. With today's high-technology systems, however, more and more training is necessary. In addition, we must all recognize that we do take the telephone for granted, and we must all guard against using it carelessly.

Abuse is considerably more serious. It occurs when someone intentionally and willfully causes the firm to incur costs for which the abuser receives the benefit. Such abuse may seem rather mild when an employee calls a friend to arrange a social engagement or transact some personal business on the company telephone (and perhaps even on company time), but it has been estimated that personal calls could account for 20 to 45 percent of an organization's daily telephone traffic. Sometimes the improved morale offered by telephone privileges for employees is recognized, however, and companies allow nearly everyone to use the phone freely. If you find yourself in this advantageous situation, you should be sure you recognize such telephone privileges as part of your pay!

Serious abuse occurs when someone makes expensive long-distance calls having nothing to do with the business and charges the call to the company's telephone number. Such abusers may not even be employed by the firm. Here are some methods that a firm may employ to secure its telephone system against such abuse. The telephone companies are very cooperative in these security efforts.

1 Locks can be installed on telephones so that the dial cannot be turned or the tone buttons cannot be accessed, and no calls can be made either inside or outside the company. In addition, there are telephone instruments that can be used only to receive calls.

2 The telephone system can be shut down after hours, when stations are unattended.
3 Both mechanical and computerized PBX systems can be installed so that all toll calls must be made through the company's switchboard. The operator must then have sufficient training and authority to control the company's outgoing telephone traffic and choose the least-costly route for each call.
4 Computerized PBX systems and some mechanical systems can be used to enable the company to restrict certain stations to calls within certain areas. This is called *toll restriction*. The extent to which outside calls can be made (interoffice calls only, local calls only, and so on) from a station is called its *class of service*.
5 Strict policies can be enforced prohibiting acceptance of collect calls and third-party billing. Refusal of third-party billing requires cooperation of the telephone company, because long-distance operators do not always request billing authorization from the third party before the call is completed. Such verifications are made on a spot-check basis only. However, some telephone companies will cooperate with customers who refuse to accept third-party charges on their bills by transferring the charges to the number from which the call was made.
6 Credit cards can be issued to personnel who are authorized to make third-party calls, so that such calls can be allocated to specific departments or individuals who are accountable for such charges and can verify that the call did involve company business.
7 Records of all long-distance calls can be kept by employees in every department, so that telephone calls can be carefully verified. Charges appearing on the monthly bill are then matched against each department's record. Such record-keeping is time consuming and expensive, but it can be justified if abuse is serious.

   The telephone company is then asked to investigate any calls that cannot be verified. Telephone companies cooperate by providing information about who received the call and if that person remembers who placed the call and what it was about. Occasionally, such an investigation reveals an error in billing, which is corrected. Charges are also removed in cases where fraud is suspected, and telephone companies are very aggressive about locating and prosecuting people who commit such fraud. When convicted, they are subject to heavy fines and imprisonment.
8 Computerized telephone-accounting equipment and services are capable of recording details of every call and producing elaborate reports of telephone usage and charges. These systems are used primarily for management information, but they also uncover abuse quite effectively. They are discussed more extensively later in the chapter.

■ CHARGES FOR TELEPHONE USAGE

Business telephone bills may be hundreds of dollars or hundreds of thousands of dollars each month. Budget allocations for telecommunications expenditures may run in the millions, especially for large companies that own their own equipment, maintain their own networks, or have very large data communications requirements. Most efforts at controlling telephone usage are directed at the monthly telephone bill.

Amounts shown on the bill are for (1) equipment and lines, (2) installation charges, (3) directory listings, (4) measured charges, and (5) taxes. Regulated carriers are required to file rate schedules, known as *tariffs*, with the Public Utilities Commission (PUC) in their states. These tariffs must be approved by the PUCs before becoming effective, and they must be available for public inspection in the telephone company offices. Separate tariffs are filed for various equipment, installation, directory, and service charges, and the composite volumes of the utility's tariffs occupy a shelf that runs the length of a room. In addition to the rate schedules, each telephone company must publish rate practices, which are administrative guidelines for the application of rate schedules. Most of the rate information that is important to telephone customers, however, is published in simpler form in the telephone directory.

☐ Charges for Equipment and Lines

Some companies now own and maintain their own equipment and pay only for the use of telephone company lines. Most companies, however, still rent equipment from the Bell System and independent telephone companies. Their monthly charges for equipment include the cost of local lines, which generally cover all local calls (calls made within the area served by the local office of the telephone company). The telephone company maintains a detailed equipment record for each customer, upon which the rental charges are based. One important reason for the trend toward ownership of telephone equipment is in the fact that these small monthly rental charges add up to significant amounts over time, amounts which are greater than the prices charged for equipment in today's competitive market.

A customer who wants to know the exact charge for a particular telephone line or item of equipment can get that information from a business service representative or by reading the tariff. It is best to obtain it from the business service representative because the charges shown in the tariff are detailed right down to the lamps under the buttons on a key telephone set, making the tariffs difficult to understand.

☐ Charges for Installation

Installation charges are one-time charges made when equipment or lines are installed, moved, or disconnected. Such charges can be significant, and a great deal of care is taken by those responsible for telephone systems to avoid unnecessary changes. One of the chief advantages of computerized telephone systems is the ease with which changes can be made, minimizing installation charges.

☐ Charges for Directory Listings

Regular listings in telephone directories are free for businesses, as well as for residents. Charges are made for unpublished numbers, boldface listings in the White Pages, and advertising in the Yellow Pages. Companies sometimes consider these costs as part of advertising expenses rather than telephone expenses.

☐ Measured Charges

Recall from our discussion on services that measured charges are based on distance in miles and the length of the call in time. Measured charges are applied to (1) local calls when the monthly service charge is for a minimum allowable number of calls, (2) local calls to nearby exchange areas (foreign exchange), and (3) long-distance calls.

Measured charges for local calls to nearby exchange areas may appear on some telephone company bills as *zone charges*. The unit of measure for computing zone charges has traditionally been called the Message Unit (MU), although Zone Usage Measurement (ZUM) is now appearing. Message unit rates are published in the White Pages of the telephone directory (usually under the heading of *nearby rates*). Each individual call appears on the bill in detail. The details may vary from telephone company to telephone company, but most include the date and time of the call, the place and number called, and the amount of the charge. The significance of message unit charges may be illustrated in this example:

> An increasingly typical charge for message units in cities which still have message units is eight cents. Let's say that your company has 100 employees and they each make ten local calls each day. This is not unusual, since studies

> show the typical employee in a large corporation to make six personal calls a day. An additional four for "business" is realistic. This works out to 1000 local calls per day. Multiply that by 22, the average number of working days in a month, and you now know that your company is making 22,000 local calls per month. Assuming that all of these calls were made in a one-message unit area, the cost would be $1760 for the month, onto which you would probably have to add local tax. However, if all those calls were to a multi-message unit area where each call is charged three message units, the cost per call is now 24¢ and the charge would be $5280 for the month, or $63,360 per year.[1]

For a major metropolitan area, this illustration is not really far-fetched. It tells why responsible managers review each month's charges for local calls carefully and try continuously to keep employees aware of telephone usage for local calls—especially the significance of personal calls.

Long-distance calls are also listed on the bill individually. In addition to showing the same details as those shown for nearby calls, the bill also shows the method of dialing, type of operator assistance (if any), and the rate period applicable. Recall from the discussion of services that long-distance calls can be made at reduced rates during nonbusiness hours and that rates for operator-assisted calls—such as collect calls and credit card calls—are slightly higher. The telephone directory shows many of these rate variations for calls made from your location to specific locations within your state, within the United States, and throughout the world.

Managers analyze charges for long-distance calls not only to detect and correct misuse and abuse, but also to seek ways of reducing long-distance costs by taking advantage of alternative services, such as WATS, private lines, and switched-voice services. If a WATS line is used, charges for it are billed separately. Switched-voice services are billed by the specialized common carrier who provides them. These separate bills are also reviewed to be sure maximum benefit is being derived from the services.

□ Directory Assistance Charges

A new charge being made by some telephone companies is for local directory assistance calls made above a certain allowable minimum each

[1]Larry A. Arredondo, *Telecommunications Management for Business and Government*, 2nd. ed. (New York: The Telecom Library, Inc., 1980), pp. 75–76.

month. For example, in New York a telephone company charges 10¢ for each request above a minimum of three requests per month.

■ GRADE OF SERVICE

Neither the telephone bill nor the logs kept by employees on outgoing calls can tell the whole story, however. Managers also need information about incoming calls to be sure service is adequate. Incoming traffic is usually studied to determine the *grade of service*, which tells how many calls get through and how many calls are blocked (encountered busy signals). For most companies, blocked calls mean lost business or poor customer service.

Blocked calls are stated as a percentage of total attempts. The result is called a *P-factor*. A company that has two blocked calls out of every one hundred attempts has a grade of service of P02. Different P-factors are suitable for different types of businesses. For example, a blocked call could be a more serious loss to an investment broker than to a service station. Most telephone companies assist with traffic studies by conducting a *busy study* free of charge to determine the number of blocked calls for a certain period. Some telephone companies also assist in determining the number of attempted calls. Generally, however, the only way the total number of attempts can be determined is by having an employee log all incoming calls. If the results of the study indicate a high P-factor, the company may decide to install more trunks or lines.

■ COMPUTERIZED TELECOMMUNICATIONS MANAGEMENT SYSTEMS

It is, of course, impossible for companies with very large communications systems handling thousands of calls to keep adequate records by hand that will provide the information management requires for planning and control. Fortunately, the computer has contributed greatly to the capability for automatic recording and processing of call detail information. The automatic systems discussed here are Automatic Identification of Outward-Dialed Calls (AIOD), Customer-Dialed Account Recording (CDAR), and Station Message Detail Recording (SMDR).

□ Automatic Identification of Outward-Dialed Calls

AIOD is a system in which information about toll calls is gathered by the telephone company. It is generally used for CENTREX systems and enables users to charge costs to individual stations.

□ Customer-Dialed Account Recording

CDAR is another system involving the phone company and requiring special equipment. Callers are required to enter accounting numbers into the system when calls are dialed. Calls are then billed to specific departments, cost centers, or individuals.

□ Station Message Detail Recording Systems

SMDR systems provide the most comprehensive information that can be made available to telecommunications managers. Companies using them report that information from SMDR reports is the basis for many

RMS II Poller furnishes management with a complete accounting of all call activity within the PBX system or company network. Photo courtesy of Teldata Systems Corporation.

cost-saving decisions. Here briefly is how SMDR works: Complete details of all calls are captured and stored on magnetic media (tape or disk) as the calls are made. This is known as Call Detail Recording (CDR). Details recorded include the station from which the call is made, date and time of the call, length of the call, place and number called, operator assistance used, and the circuit, route, or trunk used (such as FX or WATS). The data are periodically polled and processed by computer so that useful cost figures and statistics are compiled and reported in formats helpful to management. The reports can even be tailored to meet the needs of the individual company.

Companies are using SMDR systems in several different ways. Some companies attach the data collection device to their PBX systems (many are compatible with mechanical, as well as computerized, systems) and turn the data over to service bureaus, where the processing is done. Other companies install all the equipment on their premises and perform the total operation, from capture to report generation. Finally, as we saw in Chapter 3, some electronic switches have CDR/SMDR integrated into their computers.

Many of the reports and controls produced by SMDR deal with concepts you already understand, and you will recognize them in this summary of the kinds of management information generated by SMDR.

Summary Information. The system polls the stored data and adds up the total number of calls, total length of calls, and total cost of calls by division, department, or other designated cost centers. Similar polling and totalling can be performed by each individual extension or station. These totals not only help management analyze telephone costs, but also help detect abuse.

Total costs for each department or other cost center can be compared to budget allocations to help management control expenses with effective budgeting. Similarly, the actual circuit or route used and cost of the call can be compared to the least costly route and cost, and the savings that would have resulted from correct usage can be shown. This comparison may alert management to incidents of misuse. Often, however, it may mean that a station user *had* to choose a more-costly route because the ideal route was not available, an indication that more circuits may be necessary.

Exceptional calls such as calls exceeding specified lengths or costs, can be summarized. Such calls may be justified, or possible abuse can be detected.

Cost Allocations. One important accounting function is the allocation of costs so that actual expenses can be compared to actual revenues generated by a certain department or product line. Telephone expenses have always been difficult for accountants to allocate exactly. Summary information by department or other specified cost center can, therefore, enhance the accounting function.

Customer or Client Chargeback. Special account codes can be assigned to each client. The caller then keys the account code into the system and the call automatically appears on the client's next bill. A firm of attorneys would find this capability particularly useful.

Call Activity. Both incoming and outgoing calls can be monitored, although information about outgoing calls is the most useful. The number of calls placed by hour of the day and day of the month can be shown, so management can identify peak traffic periods and determine if the grade of service is adequate. Usage can also be reported by trunk or circuit to ascertain whether trunks or lines should be added or deleted. The system can also poll the numbers dialed by area code, prefix, or the entire number. The resulting report of the frequency of calls to certain locations helps determine if more FX, tie, or WATS lines are justified.

Finally, detailed information regarding the usage of FX, tie, and WATS lines is accumulated to be sure they are needed and properly used. Detailed usage reports of switched-voice services provided by specialized common carriers, such as MCI or Sprint, are also provided.

Active Functions. The reporting functions described above are all known as *passive* management functions. They simply tell what is happening. Some SMDR systems perform all these passive functions, plus additional *active* functions, which help to control costs *before* they are incurred. You will recognize some of these functions from your previous reading in this book.

1. *Least Cost Routing* As you already know, LCR means the system will automatically choose the least expensive route as the call is dialed. It is also known as *route optimization* or Automatic Route Selection (ARS).
2. *Queuing* You have learned about queuing for incoming calls. This queuing function is for outgoing calls. If the optimum route is in use, the call is automatically held in line until the proper route is available.
3. *Call Diverting* Call diverting is an alternative to queuing. Executives often feel they cannot wait for circuits to become available. The system will then

Sample SMDR Reports reproduced through the courtesy of Account-A-Call Corporation.

8110 12

CORPORATE TOTALS BY DEPARTMENTS — XYZ COMPANY — MANAGEMENT INFORMATION

CLIENT # — JUL 16-AUG 15, 1981 — PAGE

DEPARTMENT	IDENTIFICATION	CALLS #	HOURS #	MIN/CALL #	% OF TOTAL COST	TOTAL COST $	BUDGET $	BUDGET VARIANCE %	PRIOR $	BUDGET VARIANCE %
		2,766	141.8	3	2.9 %	918	883	4 % OVER	735	17 % UNDER
20-20	MANAGEMENT & ADMIN	1,118	52.5	3	0.8 %	254	240	6 % OVER	252	6 % OVER
20-25	OFFICE	2,828	98.8	2	1.2 %	391	280	40 % OVER	276	1 % UNDER
20-35	SUPERINTENDENCY	935	38.6	2	0.3 %	86	81	0 % OVER	69	14 % UNDER
20-40-50	OCCUPANCY SUPERVISE	429	21.3	3	0.2 %	48	76	37 % UNDER	48	37 % UNDER
20-55-60	PUBLICITY	204	8.4	2	0.0 %	10	19	49 % UNDER	19	3 % OVER
20-70-20	DIVISION I V VII X	16,392	669.0	2	6.3 %	2,021	1,838	10 % OVER	1,755	4 % UNDER
20-75	MERCHANDISE HANDLING	1,118	43.4	2	0.3 %	88	92	5 % UNDER	97	5 % OVER
20-80	LEASED DEPARTMENTS	1,441	67.7	3	1.4 %	440	375	17 % OVER	377	1 % OVER
93-32	OPERATIONS&PLANNING	852	91.7	6	0.5 %	161	183	12 % UNDER	184	1 % OVER
93-33	CENTRAL OPERATIONS	1,697	85.8	3	1.2 %	375	275	36 % OVER	296	8 % OVER
93-34	PROD. OPERATIONS	246	17.4	4	0.1 %	42	55	23 % UNDER	52	5 % UNDER
94-20-40	PERSONNEL	329	13.0	2	0.1 %	43	33	30 % OVER	36	10 % OVER
94-25	OFI								41	30 % OVER
94-26	ACC								20	1 % OVER
94-27	CRI								43	4 % OVER
94-30	DAT								49	2 % OVER
94-35	SUI								71	3 % OVER
95-25-30	OFFICE	392	23.0	4	0.1 %	34	31	10 % OVER	32	6 % OVER
95-70-30	CLERICAL ASSISTANCE	100	3.9	2	0.0 %	11	61	82 % UNDER	68	11 % OVER
95-75	MERCHANDISE HANDLG	1,770	94.1	3	1.4 %	444	411	8 % OVER	424	3 % OVER
96-30	CHAIN STAFF SERVICES	914	74.2	5	0.7 %	239	139	72 % OVER	145	5 % OVER
F9-20	MANAGEMENT	2,518	143.7	3	2.9 %	926	761	22 % OVER	860	13 % OVER
F9-35	SUPERINTENDENCY	1,402	63.6	3	0.4 %	125	123	1 % OVER	91	25 % UNDER

CORPORATE TOTALS BY DEPARTMENTS

Provides management with an overview of telecommunications expense for each department. Dollar amounts are listed with the "cents" removed for greater visibility. Budgetary information is provided for current month and prior month to highlight possible trends.

8110 14

HISTORICAL COMPARISON TO BUDGET — XYZ COMPANY — MANAGEMENT INFORMATION

CLIENT # — JUL 16-AUG 15, 1981 — PAGE

LEVEL	IDENTIFICATION		CURRENT	JAN 81	FEB 81	MAR 81	APR 81	MAY 81	JUN 81	JUL 81	AUG 80	SEP 80	OCT 80	NOV 80	DEC 80
20	S.F. CORP CNTL		3333	3266	3053	2929	3301	2774	3336	2893	2782	2903	3203	3349	3364
		BUDGET	3001	3305	3326	3228	3083	3094	3001	3137	2961	2723	2817	2962	3151
		% VARIANCE	11	-1	-8	-9	7	-10	11	-8	-6	7	14	13	7
93	SANTA CRUZ OPNS		578	428	396	471	552	473	535	531	455	380	354	366	419
		BUDGET	513	380	404	414	431	473	499	520	430	419	405	397	367
		% VARIANCE	13	13	-2	14	28	0	7	2	6	-9	-12	-8	14
94	MISSION OPNS		9375	6223	6856	6533	8622	6362	7599	7460	8116	8639	8464	7492	6537
		BUDGET	7140	7497	6751	6539	6537	7337	7172	7528	7508	7425	8021	8406	8198
		% VARIANCE	31	-17	2	0	32	-13	6	-1	8	16	6	-11	-20
95	BRISBANE OPNS		489	424	499	472	534	493	493	524	833	718	846	753	592
		BUDGET	503	730	590	505	465	502	500	507	577	641	726	799	772
		% VARIANCE	-3	-42	-15	-6	15	-2	-1	3	44	12	16	-6	-23
96	WALNUT OPNS		239	154	176	136	214	153	119	145	107	58	95	116	111
		BUDGET	139	108	127	147	156	176	168	162	72	82	73	87	90
		% VARIANCE	72	43	39	-7	38	-13	-29	-11	49	-29	31	34	24
F9	HOME OPNS		17069	16717	18147	15830	17176	15075	15274	15148	15340	16170	18807	17972	20087
		BUDGET	15165	18955	18259	18317	16898	17051	16027	15841	15427	15046	15573	16772	17650
		% VARIANCE	13	-12	-1	-14	2	-12	-5	-4	-1	7	21	7	14
T O T A L S			32000	27325	29371	26682	30808	25816	28782	27436	27731	29067	32045	30232	31237
		BUDGET	26462	31171	29598	29311	27793	28954	27367	27695	27092	26459	27762	29614	30448
		% VARIANCE	21	-12	-1	-9	11	-11	5	-1	2	10	15	2	3
			918	113	245	310	408	487	1426	735	98	199	276	184	127
		BUDGET	883	195	141	161	223	321	402	774	117	124	148	191	219
		% VARIANCE	4	-42	73	92	83	52	255	-5	-16	61	87	-4	-42
20-20	MA													232	308
														205	234
														13	32
20-25	OF													401	408
														356	360
														13	13
20-35	SU													120	130
		BUDGET	81	113	119	106	88	84	83	89	73	79	98	103	109
		% VARIANCE	0	-5	-32	-28	6	-4	11	-22	39	49	-8	16	19
20-40-50	OCCUPANCY SUPERVISE		48	33	24	51	40	39	142	48	23	40	32	45	24
		BUDGET	76	34	34	27	36	38	43	74	26	27	33	31	39
		% VARIANCE	-37	-2	-30	89	10	1	230	-35	-14	49	-2	43	-37

HISTORICAL COMPARISON TO BUDGET

Provides management with an overview by division and department of telecommunications budgets for a full year. Details percentage difference between amount budgeted and costs incurred. Budget is computed as moving three month average if not input.

EXCEPTION REPORT XYZ COMPANY MANAGEMENT INFORMATION

CLIENT # JUL 16-AUG 15, 1981 PAGE

CALLS OVER 15 MIN TOP 50			CALLS OVER $5 TOP 50		
MINUTES	EXTENSION	CALLED NUMBER	AMOUNT	EXTENSION	CALLED NUMBER
399.0	5109	213 466 5351	77.12	5109	213 466 5351
382.2	3190	415 838 4300	25.62	5125	415 934 2581
289.8	0923	8589	23.29	3107	703 920 7966
280.7	5125	415 934 2581	23.26	2340	212 753 4500
249.8	0923	8734	22.00	5032	5236142112
177.2	3190	800 221 8344	19.94	3190	415 838 4300
171.5	3193	82	18.84	5030	312 667 1660
156.5	5154	411	18.20	5170	503 482 9199
129.8	3241	415 957 0571	17.71	2547	206 682 0100
127.8	3255	415 421 9426	17.36	5151	513 579 7893
108.7	5000	415 441 7372	17.29	2547	206 752 1011
108.2	0923	8723	17.00	2862	605 348 8467
107.1	3246	415 957 0571	16.78	2151	213 986 2731
106.5	5131	415 836 3079	16.75	5142	602 955 7200
105.4	0923	86237	16.58	3038	203 278 7800
103.0	3241	415 957 0571	16.29	5007	212 764 2300
100.7	5172	415 968 8787	16.04	2936	312 394 2022
97.9	3241	415 957 0571	15.87	5023	206 682 6111
97.6	3107	703 920 7966	15.49	5039	212 753 4500
97.4	5125	415 944 1487	15.31	2760	602 887 2287
97.1	2641	411	14.46	5180	212 750 0400
94.4	0923	8731	14.39	5002	514 845 0115
93.7	3241	415 957 0571	14.22	5129	714 459 4081
92.7	3255	415 239 3688	14.00	2152	212 751 9340
91.9	3121	415 642 1483	14.00	5195	301 762 6076
89.2	3255	415 421 9420	13.63	2018	503 226 7811
88.9	3255	415 957 0571	13.62	2307	808 696 6767
87.7	3255	415 957 0571	13.39	2267	206 682 6111
87.3	0939	8734	13.13	2616	312 965 6690
86.8	2151	213 986 2731	13.10	2614	206 682 6111
86.3	3107	703 920 7966	13.08	5034	212 697 8977
85.1	3241	415 957 0571	12.98	3107	214 596 7981
84.5	3255	415 421 9420	12.96	2737	918 456 5053
82.6	0939	86226	12.71	2975	514 845 0115
80.7					503 482 9199
80.2					213 351 9611
79.0					213 477 9363
77.1					503 226 7811
76.9					213 477 2491
73.9					206 682 6111
73.7					209 439 4613
73.6					206 682 6111
73.4					206 682 6111
72.9					213 382 7171
72.1					212 944 8834
69.6	3003	415 824 9377	11.38	5172	415 968 8787
66.8	5035	415 931 3274	11.20	5098	702 831 2524
66.2	0923	8489273	11.20	5007	214 239 3742
66.2	3241	415 957 0571	11.19	2191	503 644 6949
66.0	5172	415 454 2464	11.11	5164	503 226 7811

EXCEPTION REPORT

Provides a list of the 50 longest calls and the 50 most expensive calls. Shows extension which placed the call. Highlights potential candidates for called number identification, potential abuse, potential off premise extension or potential tie lines.

8110 19

8110 28

WATS TOTALS BY AREA CODE XYZ COMPANY COMMUNICATIONS USAGE

WATS CLIENT # JUL 16-AUG 15, 1981 PAGE

	TOT CALL	MIN/CALL	TOT HRS	COST/CALL	TOT COST	COST/MIN
201	173	4.4	12.71	1.31	226.10	.2965
202	264	2.2	9.71	.68	180.36	.3096
203	92	3.1	4.76	.97	89.35	.3130
205	21	2.2	.78	.71	14.98	.3202
206	1,567	3.8	100.40	1.18	1,844.22	.3062
207	7	4.1	.48	1.18	8.23	.2867
208	24	3.3	1.31	.75	17.90	.2283
209	268	3.1	14.03	.61	162.67	.1933
212	813	3.9	52.39	1.25	1,013.66	.3225
213	5,292	3.2	277.91	.61	3,222.87	.1933
214	246	4.5	18.34	1.14	280.80	.2551
215	155	3.1	8.02	.98	151.21	.3142
216	68	4.7	5.32	1.38	93.61	.2932
217	11	3.1	.57	.99	10.93	.3204
218	6	12.9	1.29	3.79	22.71	.2934
219	42	4.6	3.24	1.35	56.90	.2930

AREA CODE TOTALS XYZ COMPANY

NON-WATS CLIENT # TOTALS BY AREA CODE JUL 16-AUG 15, 1981 PAGE

	TOT CALL	MIN/CALL	TOT HRS	COST/CALL	TOT COST	COST/MIN
201	33	2.4	1.35	1.13	37.41	.4636
202	108	1.4	2.45	.43	46.13	.3145
203	19	1.7	.54	.67	12.70	.3944
204	1	.4	.01			.0000
205	2	.6	.02	.33	.66	.6000
206	313	2.7	13.94	1.21	379.30	.4534
207	2	1.3	.04	.56	1.12	.4480
208	7	1.4	.16	.29	2.06	.2146
209	79	3.1	4.09	.39	30.87	.1259
211	4	1.2	.08			.0000
212	5,844	3.7	358.70	.70	4,116.52	.1913
213	1,804	1.7	51.43	.62	1,112.32	.3604
214	49	2.6	2.12	.85	41.85	.3295
215	37	2.5	1.55	1.14	42.12	.4539
216	19	4.0	1.27	.73	13.82	.1816
217	4	1.6	.10	.92	3.68	.5935
218	1	2.4	.04	1.52	1.52	.6333
219	13	1.5	.32	.54	6.96	.3683
301	175	1.6	4.58	.54	94.43	.3436
302	2	.6	.02			.0000
303	35	4.1	2.36	.41	14.22	.1003
304	5	.8	.06	.13	.66	.1737
305	29	1.4	.69	.51	14.82	.3597
306	1	.5	.01	.86	.86	1.7200
307	1	.8	.01			.0000
309	2	1.1	.04	.54	1.08	.4909
312	776	1.8	23.24	.66	509.70	.3655

WATS TOTALS BY AREA CODE/ AREA CODE TOTALS

A summary of WATS and non-WATS calls by area code. Could show the necessity for WATS service.

NUMBER OF CALLS BY HOUR BY DAY OF MONTH XYZ COMPANY COMMUNICATIONS USAGE

COMBINED CLIENT # JUL 16-AUG 15, 1981 PAGE

NUMBER OF CALLS BY HOUR BY DAY OF MONTH

Lists the number of calls by hour by day. Account-A-Call uses this report to validate the call data prior to processing. You can use this report to derive busy hour information.

	00	0	1	4	4	1	3	0	7	0	2	5	2	0	1	0	67	2	0	0	2	1	0	2	1	0	0	0	0	0	0	0
	01	0	1	2	0	1	1	0	2	3	0	1	1	0	0	2	0	0	0	0	0	2	3	0	4	0	0	2	1	0	0	0
H	02	3	0																										0	0	0	0
	03	2	0																										0	0	0	0
O	04	1	0																										0	1	0	0
	05	0	0																										1	2	0	0
U	06	1	2	3	5	2	2	11	1	1	2	9	4	10	3	0	0	1	2	0	0	7	0	0	0	1	0	6	7	4	3	0
	07	3	2	21	20	20	11	12	1	0	19	20	21	15	6	2	0	15	4	1	12	11	29	16	17	3	0	8	16	28	21	0
R	08	16	2	123	166	135	145	149	28	4	102	122	161	156	114	44	0	186	18	5	116	165	184	175	137	12	11	128	124	163	129	0
	09	108	4	399	474	538	469	508	120	3	441	497	561	496	514	130	0	482	76	5	402	491	507	497	505	96	9	443	475	491	372	5
	10	214	18	629	652	719	709	684	207	4	650	722	745	666	661	256	0	597	175	6	605	738	718	628	637	204	27	503	686	681	523	636
	11	191	26	668	673	753	700	706	185	9	642	694	706	668	673	220	0	685	137	13	677	707	705	671	721	177	25	719	639	641	618	720
O	12	176	39	571	522	481	570	561	141	27	581	554	608	568	447	151	456	547	122	45	595	506	549	515	469	114	50	589	509	531	462	549
	13	138	53	558	497	518	501	506	126	48	551	560	495	520	479	199	473	438	91	32	516	529	530	515	468	113	59	572	500	516	473	490
F	14	142	54	631	590	531	554	547	125	35	588	545	609	584	589	178	459	539	95	26	581	595	526	571	568	124	54	536	658	554	246	575
	15	134	46	566	597	643	547	568	117	44	639	639	588	543	512	153	473	513	127	27	610	531	528	532	575	120	51	573	518	548	0	540
	16	140	50	499	581	516	421	443	119	44	553	564	593	486	450	93	446	457	116	32	518	529	499	471	442	115	43	512	478	459	0	527
	17	58	17	280	290	266	358	257	65	13	318	307	268	302	258	0	324	261	61	10	267	261	257	301	202	62	9	307	282	301	0	276
D	18	11	8	125	96	48	147	102	13	10	125	64	90	119	84	0	117	80	17	10	122	35	41	129	87	4	4	156	51	54	0	89
	19	2	3	90	47	38	115	59	6	2	92	43	34	69	61	0	83	55	8	1	65	8	19	102	35	4	0	92	18	32	0	61
A	20	1	5	13	15	15	31	14	10	6	31	15	8	15	11	0	9	12	15	2	13	5	6	8	9	2	1	14	8	11	0	12
	21	2	1	10	4	5	5	10	2	4	16	1	3	5	4	0	3	14	4	2	1	1	1	5	9	2	1	3	5	4	0	5
Y	22	4	4	3	1	4	3	12	3	8	7	1	3	7	2	0	4	1	3	4	4	0	4	2	4	3	3	2	0	0	0	3
	23	3	3	5	1	1	4	3	4	2	3	1	0	0	3	1	25	6	2	1	1	1	1	4	3	2	4	3	1	3	0	5
	MONTH	8	8	8	8	8	8	8	8	8	8	8	8	8	8	8	7	7	7	7	7	7	7	7	7	7	7	7	7	7	7	7
	DAY	1	2	3	4	5	6	7	8	9	10	11	12	13	14	15	16	17	18	19	20	21	22	23	24	25	26	27	28	29	30	31

8110 31

automatically divert the call to the next least costly route. For example, if an out-of-state FX line is in use, the call may be diverted to a WATS line.

4 ***Toll Restriction*** Toll restriction is the same concept you read about previously. In this case, however, the outside calls of a certain station are restricted by the SMDR system rather than by the company's switch.

Vendors and service bureaus claim SMDR can save users from 10 to 40 percent on their telecommunication costs. Some user experience seems to bear this out. A New England bank reported savings of $15,000 as a result of a report on its WATS usage. A Boston-based university saved $67,000 as a result of an analysis of trunk traffic.[2] Using SMDR reports in conjunction with some of the controls described earlier, a Detroit hospital saved $20,000 during the first 5 months in which their system was used.[3]

Even more important, such savings can be effected without reduction in the services to users. Also, such reports can point the way to improvements in the company's communications support system. These, after all, are the ultimate goals of telecommunications management.

[2]Traudi Tissler, "Phone Usage Info and Trunk-Traffic Use Are Available From Teleprocessing Service Bureaus," *Communications News* (January 1982): 81.

[3]Daniel Gonos, "Sinai Hospital Eliminates Phone Misuse and Controls Expenses with Auditing Procedure," *Communications News* (January 1982): 57.

QUESTIONS FOR REINFORCEMENT AND DISCUSSION

1. What is the goal of telephone cost controls?
2. What are the three sources of telephone costs?
3. What are some techniques for reducing telephone misuse?
4. What are some ways for preventing telephone abuse?
5. What are the four types of charges commonly shown on a telephone bill?
6. What is meant by measured charges?
7. What is meant by grade of service?
8. Describe a telephone log for manually recording telephone charges that will later be used to verify the telephone bill.
9. Describe the following automatic telephone cost-recording method.
   a. AIOD
   b. CDAR
   c. SMDR
10. Cite some information that can be reported by SMDR and tell how it can be used for optimizing telephone expenses.

COMMUNICATION CHOICES

Choose the proper method of communications for the following messages. Some messages may be conveyed effectively by more than one method. You should decide which one you feel is optimum (provides the best possible communications support at a reasonable cost) and be prepared to explain your reasoning.

a. Visit
b. Travel for conference
c. Regular mail
d. Interoffice mail
e. Voice communication
f. Record communication
g. Data communication
h. Video communication

_______ 1. An order for urgently needed office supplies from a local vendor.

_______ 2. Regular monthly meetings of the general managers of a multinational corporation whose ten divisions are located on four different continents.

_______ 3. A discussion with a very important client who is extremely upset about a mistake made on the recently published advertising brochure sent to 6000 customers.

_______ 4. An application for employment.

_______ 5. Extensive results of medical tests analyzed by a major West Coast medical center for a patient in the critical-care unit of a Miami hospital.

_______ 6. Information requested by a regular customer.

_______ 7. The semiannual meeting of the 3-member board of directors with the six trustees of a firm doing business in the northern United States and the eastern Canadian provinces.

_______ 8. A message urging a Senator to vote no on a bill that will face balloting tomorrow.

_______ 9. A question about a letter received in this morning's mail and requiring immediate action.

_______ 10. An emergency order for spare parts for equipment that is not working at the German-based manufacturer of a small company. The parts are available only from a warehouse located in Salt Lake City, Utah.

_______ 11. An announcement regarding the appointment of a new office manager.

A CASE OF TELEPHONE TROUBLES

As president of Sims Products, a small consumer-products company with product lines of automotive parts and small appliances, Mr. Sims is very pleased with his company's recent growth. A number of newly hired management and sales personnel are expected to make the company even more successful. The firm is still small enough, however, for him to review and approve payment for all company expenditures.

This month's telephone bill showed an alarming increase. Toll calls to local exchanges increased by an even greater percentage than long-distance calls. In addition, there were a number of very expensive long-distance calls to cities where Sims Products does not have customers, so far as Mr. Sims knows. Mr. Sims feels the increase in the bill justifies his personal investigation.

He starts with the receptionist. She reminds him that she brought to his attention some time ago that the 10-line key telephone system is strained by the demands of the growing staff. She tells him that many times it is difficult for her to route calls because someone else is using the intercom line and that she just cannot find anybody where they belong. In addition, the managers yell at her occasionally when the WATS line is not available. They complain that certain people monopolize it and imply that she should have some way of controlling the problem. The new salespeople tie up all the lines around 10:30 on some mornings when they make their canvassing calls to retail outlets, and no calls can get in or out. With considerable patience, Mr. Sims persuades her to name people about whom she has heard complaints, assuring her that she is not passing along company gossip, but that he needs the information in order to improve the phone service and reduce the costs.

When he asks these people about their telephone usage, he finds that they spend a lot of time away from their desks because the intercom line is tied up and they have to track each other down to communicate. Rather than wait for the WATS line to be available, they use regular lines for long-distance calls; sometimes they don't even try to get the WATS line when they are in a hurry. Customers complain that the Sims people are very hard to reach and messages have to be left for them all the time. Their most serious complaint is that the receptionist is always buzzing them when they are on the telephone because she cannot see that they are tied up.

Mr. Sims decides to hire a telecommunications consultant for advice. What sort of recommendations do you think the consultant might make?

Technology

By now you should have a pretty firm grasp on office telephone usage. However, you may be wondering about just exactly what happens inside the wires and telephone equipment about which you have learned. This chapter gives you some nontechnical descriptions of some concepts of telephony.

■ INFORMATION

Let us begin with a clear perspective on information itself. In its broadest sense, information is defined as knowledge and intelligence, and it includes any signals or symbols recorded or transmitted with intended meaning. For the people involved with the communications activities under discussion here, information is defined as a signal purposely impressed upon the input of a communications system.

In telecommunications, all information and intended meaning is conveyed with electric signals impressed upon the input of a communications system. Here are some examples.

- When you lift your receiver, you send an electric signal to the exchange office of your telephone company that says you want to be connected with another subscriber on the system.
- When you turn the telephone dial, you send a series of electric pulses to your switchboard or to the local exchange office of the telephone company. The signal you send specifies the "address" of the location on the system with which you wish to be connected. If you have a Touch-Tone telephone, you address your call by sending a series of electronically created harmonic tones.
- When you wish your grandmother a happy birthday, make elaborate social plans with a friend, or transact business with another company, the sound of your voice is carried as an electric current.
- When you hang up by replacing your receiver on the switchhook (or by turning a switch on terminal equipment to an off position), you stop the electric current from flowing and tell the exchange that you no longer want to use the system.
- In data communications systems, a stream of electronic pulses representing coded numbers, symbols, or letters, flows along a transmission path just like boxcars on a train, but at speeds beyond your perception.
- In video communications, the components of a sophisticated electronic system pick up a whole scene, sometimes with live action, convert it into a stream of electronic signals that can be transmitted over telecommunication channels, and reconstruct it on a screen at the receiving station.

Everything transmitted, whether it is a simple signal instructing a switch, a conversation between friends, a very long stream of highly confidential, encrypted data, or a video conference, can be considered information. In fact, information today is recognized as a vital resource in business and government.

■ SOUND

Most telephone transmisssions are voice communications in which information is transmitted in the form of sound. Sound is caused by the vibration of a solid body; this vibration creates wave motions in the air that radiate from the vibrating body, which is the source of the sound. The radiation is often compared to the waves seen on the surface of water that has been struck by a pebble. Different sounds are created by variations in the frequency of the vibrations and, in turn, the variations of the sound waves. Speech and harmonic sounds are created by complex waveforms with several sequences within each cycle, or several little waves within each big one. The sound is measured in complete cycles per second, or *hertz* (Hz). The volume or loudness of the sound is determined by the *amplitude* of the sound wave, which is the amount of energy in it. Volume is measured in *decibels* (dB).

■ SPEECH

Recall from our model of human communication that ideas are transmitted from one mind to another by way of speech. The speech process begins when the idea is encoded into verbal symbols by the sender. These verbal symbols are sounds. The vocal chords of the sender are the source of the sounds, and the carrier of the sounds is the air over which the waveforms travel. The receiving mechanism is the ear of the listener, which picks up the sound waves and transmits them to the brain of the receiver for decoding. If all goes well through the process of encoding, conversion to sound, transmission through the air, reception, and decoding, exactly the same idea will exist in both minds, and communication has occurred.

■ TELEPHONY

Air, however, will carry sound waves only a limited distance. Energy is lost along the way, so that the sound becomes fainter and fainter until it can no longer be heard. The purpose of telephony is to convert the sound waves coming from the sender's vocal chords into electric signals, carry them for long distances over wires or radio beams, and then turn them back into sound waves in the air near the receiver.

The conversions from sound waves to electricity at the sender's station and then from electricity back to sound waves at the receiver's station are accomplished in the telephone set itself. In ordinary telephones, the handset contains both the transmitter and the receiver.

The microphone in the mouthpiece of the sender's handset converts the sound waves into electrical signals. The microphone has an iron diaphragm, which vibrates when it is hit by the sound waves. The diaphragm then communicates those vibrations to some carbon granules, which create an electric current with exactly the same frequency as the vibrations of the original sound. The electric current with the frequency matching the sound waves is combined with additional current that will carry it along its path to its destination. We will say more about the carrier paths in just a moment.

When the current reaches its destination, it encounters a permanent magnet and an electromagnet in the earpiece of the receiver's handset. These two magnets control another diaphragm, causing it to vibrate with the same frequency of the current it receives, and then to emit into the air the very same sound waves as the vocal chords of the sender emitted.

In telephone jargon, the encoding and decoding processes, which convert the sound waves to electric signals and back again inside your telephone handset, are called *modulating* and *demodulating*, respectively. In Chapter 5, in our discussion of encoding and decoding for data communications, the device that acccomplishes such conversions is called a *modem*. The transmitter and receiver in your telephone handset are also modems. The human brain, in fact, may even be thought of as a modem when it converts ideas into verbal symbols and sounds. In this chapter, you learn that there are different modems for different purposes.

■ CARRIERS AND CARRIER SYSTEMS

Until now, the word *carrier* has been used in only one context—naming the companies that provide carrier services to users and subscribers. Now we are going to add another meaning to this important verbal symbol. Carrier also denotes the path over which telecommunications travel. Additional names for telecommunication paths are *line, circuit, trunk, channel, link, band,* and *beam*. You learned in Chapter 5 that the various telecommunication carriers include wires, cables, and radio beams. Most of these carriers can handle several conversations or transmissions simultaneously. When they do, they are called *carrier systems*. There are several facts you should know about carrier systems.

□ Mode

In the simplest form of transmission, information is carried in only one direction. This is known as the *simplex* (SPX) mode of transmission, and an example of it is the signal you send to the telephone office when you lift your receiver off the switchhook. The information conveyed by the signal says you want to use the communications system. Another, more-commonly used mode of transmission is called *half-duplex* (HDX), which means transmission can travel in both directions on the same line, but not at the same time. An example of the half-duplex mode is a single conversation on a tie line—one person talks and then the other person responds, but both people are never transmitting at the same time. Another example of HDX is the link between your house and the telephone line that runs to the local exchange. Both SPX and HDX require one pair of telephone wires. When carriers are handling many conversations or transmissions simultaneously, however, transmission is flowing in both directions at once, and the mode is called *full-duplex* (FDX). Full-duplex requires a carrier of four wires, or two pairs. It is important to note that the FDX mode makes the most efficient use of the carrier.

□ Analog Transmission

The difference between the ordinary electric current that turns on your coffeepot and the electric current that flows into your telephone is that the telephone line carries information in the form of an *intelligence signal*. The intelligence signal is the part of the current that matches the frequency variations of the sound waves. It travels like a wave, superimposed on the carrier current. This continuous-wave intelligence signal is known as *analog* transmission.

As the combined current flows along the line, some energy loss occurs, just as sound waves lose power in the air. This amplitude loss is called *attenuation*, and it results in lower volume. Therefore, it is necessary to have *repeater* devices along the way to *amplify* the carrier current in the wires and keep the signal strong. In the ordinary telephone wires running to your house or office, repeaters may be required about every 6000 feet. Different types of carriers require different kinds of repeaters at different intervals. In microwave radio transmission, the microwave relay stations, which look like giant dishes, are the repeaters, and they are required every 30 to 200 miles, depending on whether the signal is sent in a straight line or along the curve of the earth. Other devices work to control the current in the line (electricity tries to "get away") so that noise, distortion, and interference are reduced and crosstalk among channels is avoided. When the current reaches its final destination, the intelligence signal is separated from the carrier current by a device known as a *filter*. Much of this equipment is underground in the manholes in which you see the telephone maintenance people working.

□ Digital Transmission

The continuous analog waveform is not the only intelligence signal. Another method of carrying such signals is *digital* transmission, which is the method you learned about in our discussion of data communications in Chapter 5. Recall how in data communications, every number

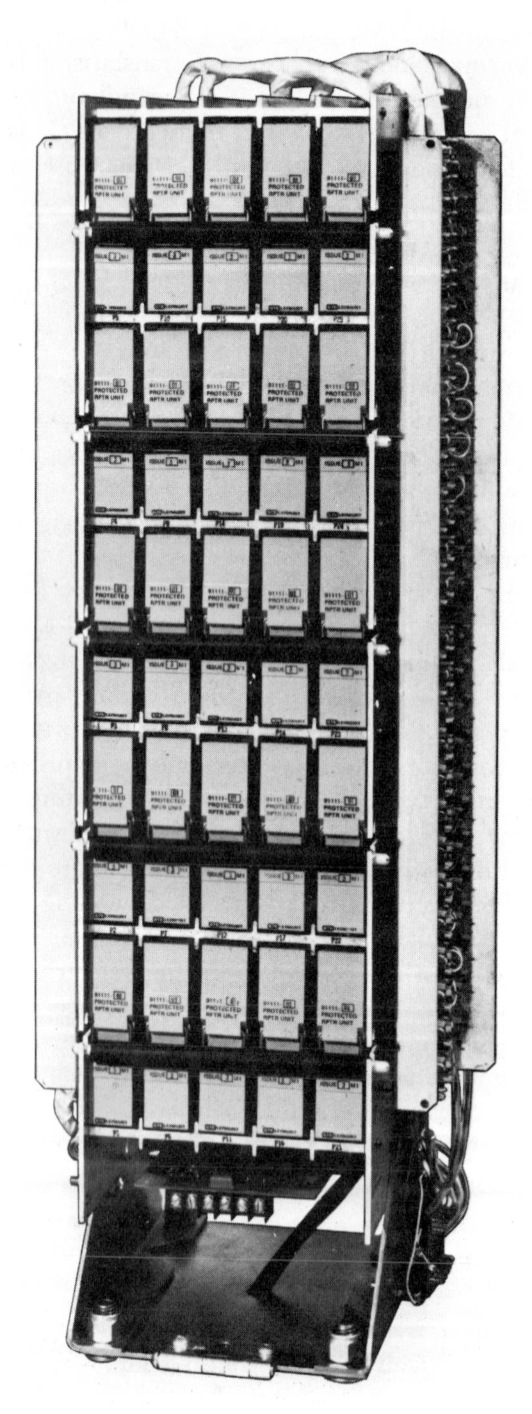

(a) 9101C repeater nest.

(b) Repeater housing. Photos courtesy of GTE Lenkurt.

and symbol is translated into a code and represented by some combination of the two digits 0 and 1. These 0s and 1s are actually on and off pulses. Each pulse is called a *bit* and the transmission rate is stated in bits per second. Just as data are converted into the transmission code, the voice signal can be digitized by converting the analog signal into a bit stream. One method of converting the analog wave into a bit stream is called *Pulse Code Modulation* (PCM). In PCM, numeric samples of the frequencies of the sound waves are converted into electronic pulses, which form the bit stream. Because the digitized signal is a series of pulses rather than a continuous wave, noise and interference are not transmitted or amplified by repeaters, which makes digital transmission highly desirable. However, some control devices and maintenance are still required for smooth operation of the communication system.

Another advantage of digital transmission is increased carrier capacity and speed. Modern carrier equipment can handle millions of bits per second. Any type of communication (voice, data, and image) can be converted to a digital signal. In addition, any type of carrier (wire, cable, or microwave radio beam) can be equipped to transmit digital signals. In fact, experts predict that nearly all transmission of the future will be digital, and that the circuits of the future will combine bit streams and carry all types of communications simultaneously.

□ Multiplexing

The techniques that allow a carrier system to handle several conversations or transmissions simultaneously are already in use and have been for some time. They are called *multiplexing techniques*. The importance of multiplexing can easily be understood when you try to imagine how many wires would be required to handle all of today's traffic if each pair of wires could handle only one conversation. The two multiplexing techniques discussed here are Frequency Division Multiplexing (FDM) and Time Division Multiplexing (TDM).

□ Frequency Division Multiplexing

Frequencies of sound waves are measured in hertz (Hz), which represent the number of cycles per second in the sound wave. The sounds created by the human voice are limited to the range of 100 to 800 Hz. The sounds that can be picked up by the human ear range from 20 to

20,000 Hz. (As a point of reference, high-fidelity music requires a range of 30 to 20,000 Hz, and the sound of middle C on the musical scale is exactly 200 Hz.) The frequencies used to carry telephone conversations range from 200 to 3400 Hz, because this range is all that is required for accurate voice communication, even though human speech and ears may use a wider range.

Fortunately, even the smallest copper wires used in the telephone systems of the world today can be made to accommodate a wider range of frequencies than 200 to 3400 Hz. With FDM the conversations or transmissions are stacked on the channel in different frequency ranges. For example, a cable carrier may have a capacity for carrying frequencies of 100,000 Hz. Your voice occupies only 200 to 3400 Hz on that frequency spectrum, leaving the remainder of the spectrum available for other conversations. The total range of frequencies on the carrier is the *bandwidth*, and the range occupied by your conversation is the *bandpass*. Your conversation may be modulated upward and carried at a bandpass range of 4200 to 7400 Hz, while your neighbor's conversation is carried at the normal 200 to 3400 range.

FDM is the most frequently used carrier system for voice channels. Ordinary telephone wires and cables may be multiplexed to carry from 1 to 24 conversations. Coaxial cables in the L series of AT&T are organized into groups called *supergroups, mastergroups,* and *jumbogroups*, carrying 60 voice channels, 600 voice channels, and 3600 voice channels, respectively. The L5 coaxial cable has a capacity for carrying 10,000 voice channels. Bandwidths for these high-capacity channels are stated in kilohertz (kHz, thousands of hertz), megahertz (MHz, millions of hertz), and gigahertz (GHz, billions of hertz).

Microwave radio channels carry from 1200 to 1800 conversations. A recently developed carrier, known as a *helical waveguide*, carries very high radio frequencies in an empty tube about the size of your wrist with a capacity for 230,000 simultaneous telephone conversations.

□ Time Division Multiplexing

TDM is based on a totally different principle. Carrier systems using TDM interleave conversations and transmissions on the carrier in different time slots, rather than on different frequencies. This is possible because (1) there is a lot of empty transmission space between your words, sentences, and responses, and (2) transmission can occur faster than you can talk. All this extra time can be made available for

9120A digital multiplexer. Photo courtesy of GTE Lenkurt.

other transmissions to occur on the same channel. Moreover, all forms of communications (voice, data, and image) can be digitized, multiplexed, and transmitted over the same high-capacity, high-speed channel.

The devices that accomplish both FDM and TDM are called *multiplexers* and *demultiplexers*. These different multiplexed carrier systems, along with their varied capacities, bandwidths, and transmission speeds, now tie the world together with a network of wires, cables, and radio links, over which vast quantities of information travel.

■ SWITCHING

You may wonder how these signals get to the right destination in this vast network. The answer lies in switching. *Switching* is the term used to name the interconnection process of the communications system, whether it is voice, data, or image. The success of any carrier network is dependent upon its ability to connect any station or terminal equipment to any other station or terminal equipment rapidly and accurately. The switching functions are much more complicated today than they were when Theodore Vail envisioned his grand telephonic system.

If you had made a telephone call in 1880, you would have lifted a receiver off a switchhook and turned a crank on the side of the telephone box. The raising of the switchhook would have opened your line, and the turning of the crank would have sent an electric current to the

Boy operators at a Louisville, Kentucky, telephone exchange around 1900. Photo reproduced with permission of AT&T.

office of the telephone company strong enough to light a lamp on a switchboard. When the human operator saw the lamp glowing, he or she (young boys were employed to operate switchboards in the beginning, but they proved unruly and were replaced by women) would have plugged a cord into a jack connected to your line and asked you what connection you wanted. (Today you get a dial tone telling you that the system is ready to serve you.) There were no telephone numbers then, so you would have spoken into the transmitter on the box and simply said the name of the subscriber you wanted. The operator would then have found the jack for that subscriber, tested the line to see if it was in use, and either told you the line was busy or plugged another cord into the jack to connect your two lines. This would have activated a ringer in the telephone box of the person you called. When that person lifted the receiver, the connection was complete, and your lamp would have gone out. When you were finished with your conversation, you both would have hung up your receivers and your lamps would have glowed. When the operator noticed this condition, he or she would have removed the plugs from both jacks, the lamps would have stopped glowing, and the system would be ready for the next time you wanted to use it. As a final step, the operator would have recorded the charges for the call.

A few small telephone companies in rural areas and some private exchanges in small offices are still operated in this manner (but usually without the crank). However, it has long been recognized that if we were still dependent upon manual switching today, there would not be enough women in the world to handle all the calls!

In addition, toll and long-distance calls made in today's worldwide networks must do much more than just find their way along a local loop to the exchange office. They must also travel along trunks from exchange office to exchange office, through toll centers and along cables and microwave links of the DDD network, through the networks of SCCs, or perhaps even into the international network. Calls that are placed through a private computerized switch that controls a corporate network first find the least costly route and then may be channeled either onto the corporation's private network or onto a public carrier's network. Indeed, a call from Los Angeles to Cincinnati placed through a corporate switch may be more-cheaply routed on the corporation's private line to New York and then back to Cincinnati on the public network than from Los Angeles directly to Cincinnati on the public carrier's network—and it would get there just as quickly.

Because today's automatic switching equipment can route calls so rapidly and because public circuits are so heavy with traffic during peak periods, calls may travel on whatever circuit is open, regardless of whether or not it is the most direct route. During holidays, a call

Wall telephones with cranks, used from 1907 to 1917. Photo reproduced with permission of AT&T.

made from Chicago to Denver may find available routes only through Miami. Because peak periods do not occur simultaneously in different time zones, a call made from New York to Washington, D.C., may be routed through California! Regardless of the route used on the public network, however, the switching can be accomplished in seconds and the cost of the call is the same.

Automatic switching equipment employed by the public carriers to accomplish all this may be divided into the same two categories as the switching equipment described in the section on private branch exchanges in Chapter 4. They are (1) mechanical switching systems, which include step-by-step, panel, and crossbar switching methods, and (2) electronic switching systems, which are computerized switches.

□ Electromechanical Switches

The first electromechanical switching system was developed in 1889 by Almon Strowger, and updated versions of it are still in use today. The Strowger switch introduced both the rotary dialing method for placing calls, which meant every subscriber had to be assigned a number, and the step-by-step switching method. The step-by-step switch has a vertical rod with a metal arm attached to it. When the rotary dial is turned,

(a) Step-by-step switch.

(b) Crossbar switch.

the pulses sent to the switch cause the arm to go step-by-step through an array of contact terminals arranged in a circle around the rod. If you are calling our fictitious Sims Products at 555-1674 and you dial the number 1 in the thousands place, the pulse you send causes the arm to step up one place in the fourth row of the array. When you dial the 6 in the hundreds place, the arm steps six places in the third row of the array, and so on. This stepping action occurs as you turn the dial, which means that you have control of the switch during the entire dialing and routing process. Therefore, this switching method is slow and cumbersome compared to switches developed later.

A second switching method also used today was introduced in the 1920s; it is called the *panel-switching method*. Panel switching is faster than step-by-step switching for two important reasons. First, it can store the number dialed and release the switch for the next call sooner. Second, this switch has more terminals, which results in shorter connection time when circuits must be searched for availability. A typical

(c) First channel switch. Photos reproduced with permission of AT&T.

panel switch has 500 terminals arranged in 5 rows of 100 each on a large frame or panel. A selector rod searches the panel to locate the dialed number and select the route. A similar switching method, called the *rotary method*, operates on the same principle as the panel switch, except that the selectors move in a circular motion rather than in a straight-line motion.

A third type of electromechanical switch—and the most-sophisticated one in use today—is the crossbar-switching method. The crossbar switch was introduced in 1938, and an improved version of it was developed in 1947. It is faster than either the step-by-step or the panel switch because its rectangular sets of horizontal and vertical bars are arranged so close together in a grid formation that connections can be made in a split second. Like the panel switch, the crossbar switch can store the number dialed while the connection is being made. It is also compact, due to the close arrangement of the crossbars, and therefore occupies less space. The crossbar switch can handle thousands of calls per hour, which is adequate for serving many of the nation's small cities or less-densely populated areas.

All these electromechanical switches perform the other tasks of the operator as well. They constantly scan the system and find the line that is signaling it wants to use the system. They send a dial tone when the system is ready for use. They send a busy signal when the station called is in use, or activate the ringer in the telephone of the station called when it is available. They even record the time and charges for the call. However, operations performed mechanically are always slower than the same operations performed electronically, so electronic switching systems had to be developed to meet the high volume of telephone traffic of the 1970s and 1980s and anticipated in the future.

☐ Electronic Switching

The electronic switches used by telephone companies operate on the same principle as those on equipment described in Chapter 4. They are, of course, much larger, and some of the functions are different because the carrier's switch serves the public, while the private switch serves only the company that owns it. Basically, an electronic switch is a computer whose central processing unit performs all the tasks of the human operator at the manual switchboard. This computer has two memories. One is a short-term memory for scanning the system and recording changes. The changes it finds might include a signal that a user has

(a) AT&T's Electronic Switching System (ESS), located in Trenton, New Jersey.

(b) AT&T's ESS No. 4 network status board in Nashville, Tennessee. Photos reproduced with permission of AT&T.

accessed the system by lifting the switchhook, the number the user has dialed, when circuits are in use and when circuits become available, and so on. The other memory is a long-term memory, which contains the instructions that tell the computer what to do about those changes (such as sending a dial tone telling the user that the system will accept instructions or choosing the correct available circuit). The electronic switch does all this so rapidly that it appears to be handling several calls at once. A very important function of the electronic switch is the automatic recording of charges. (Station Message Detail Recording is really nothing new—telephone companies have been doing it ever since the first exchange opened in 1877. It is only recently that the capability for doing it automatically has been available to the business user.)

The primary advantage of electronic switches is their capacity for handling traffic in very large metropolitan areas. AT&T's No. 1 ESS can serve 96,000 terminal stations (telephones) and handle a maximum number of 105,000 calls per hour. GTE Automatic Electric Company's No. 1 EAX serves 30,000 terminals and handles up to 79,000 calls per hour, and ITT's METACONTA L can serve 64,000 stations with 100,000 calls per hour.

■ OTHER TERMINAL EQUIPMENT

Thus far in our discussion of technology, we have concentrated on the ways in which voices are carried and switched on the communications web that wraps the world. In voice communications, the terminal equipment is the telephone and people do the talking. However, as you know, there are three other forms of communications—data, FAX, and video. People are not the only communicators; machines are talking too. These machines are also considered terminal equipment.

Some machine communications are carried and switched over the same networks as voice, and some are carried on channels for machines only. Some channels merely connect two locations; they are called *point-to-point* links. Some connect a central location with several remote locations and are called *point-to-multipoint* networks. Others are *switched networks*, permitting communication between any two points on the system.

The machines that are doing the talking are teleprinters used for TWX, Telex, and data communication systems; computer terminals; facsimile machines; word and information processors; and television cameras. Often these machines use different codes and protocols and require translators in order to communicate, just like people of different nations who speak different languages. Such machines are said to be *incompatible*, and require special interface equipment or software to enable them to communicate.

Envax 500 communications system works with most word processors, any ASCII terminal, TWX, Telex, DDD, and other record carriers. Photo courtesy of ENVAX Corporation.

There are several good reasons why incompatible machines and networks have grown up. One is because manufacturers develop different office machines to meet totally different needs, many of which involve tasks other than communications. Another is because different communications systems were developed and built during different stages in the technological history of communications. The capital investment in a carrier system and its auxiliary equipment is so great that systems cannot be scrapped simply because they become obsolete. So long as a system performs a productive function, it will remain in use. Communication systems tend to grow in layers.

It really all began with the telegraph machine that Samuel Finley Breese Morse built at the Vail family's Speedwell Iron Works in 1838, nearly 40 years before Alexander Graham Bell invented the telephone. Morse code, which was transmitted over an electrical circuit, is an on-off pulse code somewhat like the machine codes in use today. The speed for early telegraph transmission was about 20 words per minute over narrowband channels with capacities below those of today's voice channels.

Teleprinter exchanges also flourished along the way. The two public teleprinter exchanges, Telex and TWX, differ from each other both in the codes used and in the carrier systems employed.

Telex still transmits over the same narrowband carriers used for telegraph transmission. Domestic Telex uses the 5-bit Baudot code, named after its inventor, Emil Baudot. International Telex uses a similar 5-bit code, known as CCITT #2 because it is approved by the Consultative Committee for International Telephone and Telegraph.

TWX is transmitted on voiceband channels, largely because it was at one time owned by AT&T. TWX uses a 7-bit ASCII code that is frequently used for data communications.

Older teleprinters produce the coded information on paper tape, with a hole punched for the on position and no hole for the off position. Transmission equipment then sends an electric pulse for each hole it reads. Modern teleprinters eliminate the tape and encode the information electronically. In fact, today's intelligent teleprinters have storage and editing capabilities similar to word processors. Some are compatible with all three teleprinter exchanges, codes, and carrier systems.

Veritable libraries have been written on today's data communications technology and the wide variety of terminals, codes, protocols, interfaces, and transmission techniques in use and under development. Input for some terminals is accomplished by voice or Optical Character

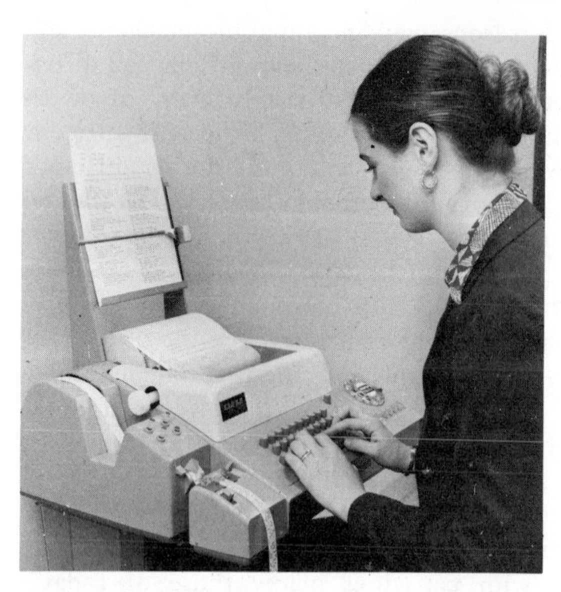

Telex teleprinter, showing punched paper tape. Photo courtesy of Western Union.

Reader (OCR) rather than keystroking. Terminal equipment for facsimile and video communications accept input from light sources.

With the advent of the integrated electronic office, more and more communications systems are demanding compatibility among these various pieces of terminal equipment. As we enter the Information Age and plunge into the future, technology is indeed meeting these demands for compatibility, and a remarkable condition of total communication is beginning to change nearly every phase of our lives. As James Martin, well-known computer systems and telecommunications author, stated in his book entitled *Future Developments in Telecommunications*, "Almost every means of communication known to man, except love-at-first-sight, can be converted to an electronic form." With proper management, the effect of such total communication can be expected both to increase productivity in offices and to improve the quality of work-life for knowledge workers.

PARAGRAPHS FOR REVIEW

Test your understanding of telephony by inserting the proper word or words in the blanks found in the paragraphs below. Choose from the words given in the lists; there is a separate list for each paragraph. You should use each word once.

1. Sound and Speech

Sound is created by __________ in the air. These motions __________ from a vibrating source. In human communication by speech, the source of the sounds is the __________ of the speaker. The tone of the sound is determined by the

__________ of the vibrations and stated as the number of cycles per second, a measure known as __________. Speech sounds are complex __________, with smaller sequences of vibrations within each cycle. Communications occur when the waveforms travel through the air from the vibrating vocal chords of the __________, are picked up by the ear of the __________, and vibrate the __________ of the listener with the same frequencies as those emitted by the vocal chords of the speaker.

eardrum	listener	vocal chords
frequency	radiate	wave motions
hertz (Hz)	speaker	waveforms

2. The Telephone

The telephone is a device for __________. Waveforms can be made to travel as an __________ over wires or __________ for much greater distances than they can travel through the air. The __________ in the telephone handset converts the soundwaves into an electric current with the same frequencies as the sound emitted by the vocal chords of the speaker. This electric current is known as an __________. The transmitter is called a __________. Another electric current, known as a __________, is added to the intelligence signal to push it along the transmission path. In another modem in the handset of the receiving station, the electric current vibrates a __________, which emits sound waves matching those of the speaker into the air around the listener.

carrier current	intelligence signal
diaphragm	modem
electric current	radio beams
far speaking	transmitter

3. Analog and Digital Transmission

Most telephone voice communication travels as an electric current known as an __________, which is a __________ matching the variations in the frequencies of the sound vibrations created by speech. Some voice communication travels as a __________, however, which means that the frequencies are converted into a series of __________ pulses known as a __________. One way of converting analog signals into digital signals is by __________, in which the frequencies of the analog signal are sampled, given numerical values, and then converted into a binary code for transmission. Because of technological advantages, experts predict that the transmission methods of the future will be all digital and that all four forms of communication (__________, __________, __________, and __________) will be transmitted simultaneously on the same channels.

analog signal	FAX
bit stream	on-off
continuous wave	pulse code modulation
data	video
digital signal	voice

4. Carrier Systems

The methods for achieving simultaneous transmission on a single carrier system are known as ________. One technique for multiplexing is ________, in which transmissions or conversations are stacked on the frequency bands of the carrier. The total capacity of the carrier is called the ________, while the frequency range occupied by a specific transmission or conversation is called the ________. ________ can carry thousands of conversations simultaneously.

Another multiplexing technique is known as ________ because different transmissions are broken into pieces and ________ into different time slots on the carrier. When the pieces reach their destination, they are put back together by a ________.

bandpass
bandwidth
coaxial cables
frequency division multiplexing
interleaved
demultiplexer
multiplexing
time division multiplexing

5. Switching Methods

One of the most important services of a carrier network is ________, which is the carrier's ability to connect any two stations (or more than two for conference calls), quickly and accurately. Originally, switching was accomplished ________ by operators who plugged cords into holes where subscriber lines terminated. As telephone traffic increased, it was necessary to develop ________ switches, which operated automatically. The three most common types of mechanical switches still in use today are ________, ________, and ________ switches. The fastest and most-sophisticated switching systems are ________ switches, which are actually ________. These computerized switches have two memories, one that is a ________ storage for recording all of the changes that occur in the system, and one that is a ________ storage containing the instructions for what to do about those changes. For example, when a ________ is lifted, the temporary memory records the signal that a subscriber wants to use the system, and the permanent memory instructs the system to respond with a ________.

crossbar
dial tone
computers
electronic
manually
mechanical
panel
permanent
step-by-step
switchhook
switching
temporary

6. The Integrated Electronic Office

When machines talk to each other they use binary codes in which every letter, character and symbol is represented by a different combination of 0s and 1s (or on and off pulses). Each of these signals is called a ________. Transmission speeds in the millions of ________ can be achieved. Three examples of binary codes are the ________, a 5-bit code used for domestic Telex transmission; the ________, a similar 5-bit code used for international Telex transmission, and the 7-bit ___________, which is used for TWX as well as some data transmission. When different machines use different codes, they are said to be ___________. One very important trend in office technologies is the ___________ of automated equipment. Incompatible office equipment can communicate with the aid of ___________. Telecommunications networks that enable office machines to communicate are expected to help increase ___________ in offices.

ASCII code	bits per second	incompatible
Baudot code	CCITT #2 code	interconnection
bit	software	productivity

PROJECT

Draw a diagram of human communications by telephone. Use the model presented for human communications in Chapter 7 and add the extra step of transmission of voice by telephony. For more examples of telephone transmission, consult any modern encyclopedia under telephone *and* telecommunications.

The Future

Mr. Sims, Sr., passed away quite suddenly in 1991, and the company was thrust into the hands of Mr. Sims, Jr. Junior had purchased a personal computer in 1981 and had been totally distracted by home computer applications and cable TV entertainment ever since. He was one of the first subscribers to a major teleshopping service and a videotex newspaper, and he had a reputation as a fanatic about using his EFT (Electronic Funds Transfer) card instead of using money or checks. There was a great deal of concern that he played with computers all the time and could hardly know enough to manage Sims Products, which had now grown considerably.

Much to the surprise of everyone, however, he quickly raised enough capital to enter Sims into a joint venture involving the manufacture of a totally new product line in space. The name of the new company was Off-Planet Products. On October 3, 2004, OPP became the fifth sharing member in the very large orbiting facility for manufacturing operations that had been launched and assembled in 1995 as a joint effort of two American conglomerates and the French government. The products planned for OPP were primarily memory and logic components for electronic toys.

It had taken $1\frac{1}{2}$ years to staff the new company and establish new information systems for coordinating the off-planet operation with the headquarters office in Denver and the two existing divisions, which were now spread out over the entire United States. Fortunately, the new venture required high-capacity digital links for interactive video communications. When the network was in place, Sims was finally able to integrate its entire automated office with all its branches and agents located throughout North America and Europe, which was a goal Mr. Sims, Sr., had established for the company shortly before his death.

The new employees who were hired to reside in the orbiting facility were regarded as pioneers and given a highly publicized send-off when they were shuttled to their new home by a Columbia spaceship. There was a tremendous celebration when the first products came off the line, and each Sims employee was given a talking doll containing one of the memory chips manufactured in space as a memento of the occasion. The dolls were capable of carrying on short conversations in three languages. All were impressed, but the concept of off-planet production was still very new and faced with risk and uncertainty, and some Sims employees regarded the whole thing with a great deal of suspicion.

Many others saw potential in the venture. After all, Junior had been right about his teleshopping and videotex newspaper. By the turn of the century, most of the homes in North America had been equipped with interactive television terminals. The concept had taken hold quite rapidly because most people already had both telephones and television sets, and the cable was widely installed by the end of the 1980s. Most shopping was now done from the home by videophone. Selections could be viewed on the screen and orders entered through a keyboard under the screen. Charges were transmitted directly to the banks through the regular EFT system.

It had long since become exorbitantly expensive to print news on paper and deliver it by truck or auto to individual homes. Both paper and gasoline had become extremely scarce before the year 2000, and only a few highly specialized magazines were printed. Nearly everyone was viewing the morning news on the screen by calling up only the sections they wanted to read. Classified ads and yellow page listings could be searched in the same manner.

Junior was right about his EFT card, too. Paper money and checks were almost nonexistent, and all transactions were transmitted through point-of-sale (POS) systems feeding into several regional computer centers accessed by all the major banks. No one ever handled money. The resulting reduction in crimes such as muggings, robberies, and burglaries exceeded everyone's expectations. Many felt it was the reason that the crusaders for gun control finally faded away.

All things considered, Mr. Sims, Jr., turned out to be right about home computer services, so the chances were very good that Sims Products would be a successful pioneer in off-planet manufacturing.

Ellen Clausen had joined Sims as a word processing machine operator in 1998, just after she completed secretarial school. In 2005, 7 years later, she enjoyed a very responsible administrative position on a team of highly competent knowledge workers. She had programmed her doll to say "Good morning, Sims Products. May I help you?" in English, French, and German. Her areas of specialization included publications and records. In this capacity, she was familiar with all of Sims' information processing equipment and had frequent contact with every division manager.

She was a bit awed by the idea of a 5-mile industrial complex in space, but she was pleased when her friend Cindy joined the staff in orbit so she would have extensive interface with the new operation as well. One of the video channels was usually open around the noon hour, and it was fun to have lunch with a friend via satellite. The space complex received all the regular video entertainment channels, so they could discuss last night's play or just catch up on family news. When they tired of Pac-Man, they rediscovered some old-fashioned card games, such as bridge and gin rummy.

Ellen lived in one of the new country developments 60 miles from the outlying suburbs of Philadelphia. She and her husband, John, were glad their children were growing up with the advantages of both the city and the country. Public-transit systems to the inner cities played an important role in the success of the country developments, but it was the public services such as teleshopping and telemedicine that really made them ideal.

The public school the children attended had only 80 students in grades one through eight, but it had all the advantages of the best teachers and programs by way of video links. Today would be a special day for them—a tour of Kyoto by satellite, conducted by one of Japan's leading historians.

John was a systems analyst with a major data base management service, and both parents were able to do enough of their work at the terminal in their home so that one of them was with the children all the time when they were not in school.

Ellen set the home security device to interconnect with the local patrol office as soon as the children left for school. Today she would have to work from the remote terminal Sims maintained with a time-share service about halfway between her home and the Sims downtown office. Mr. Sims had requested some information from an old auto parts file running from 1980 to the present. New software was being developed for auto parts, even though that division was being cut back due to the fact that the automobile was no longer the most important means of transportation.

She was miffed that she had to interrupt her other tasks to do it, primarily because it was on an old microfilm media and the reader caused severe eyestrain. Ellen had participated in the series of studies in 1999 that resulted in important ergonomic improvements in those CRT screens. At least she wouldn't have to do the updating, which used to give her headaches. Her thoughts took her back to the early days of information processing when the tasks of knowledge workers were still dull and repetitive and equipment and procedures had not yet been developed to fit the needs of people. There was, of course, still room for improvement.

The train whisked her to the location of the time-share building in a quarter of an hour, and the sound of the sensor recognizing her handprint brought her back to the present. As the automatic security doors opened, she looked forward to assembling the data for the new appliance cost lists. The appliance costing system was the most advanced of its kind. All the input could be done by voice recognition, and nothing had to be keyed. Subsequent changes could be made by remote voice input from almost any telephone, which would make the operators happy. As she logged onto the system and notified the message center of her availability, she realized that it was a good idea to get the auto parts system modernized too. Everyone seemed to underestimate Mr. Sims, and she was guilty, too.

A familiar aroma of hot coffee wafted from a hallway. Offices would never change. Ellen felt at home with the sounds of electronic equipment in operation. She often mused that those machines sounded more like gently shifting sands than the lightening-fast information carriers they were. She was chatting with some of the other early arrivals when Alan Green came in. She and Alan had gone to college together, and she had not seen him for a year. He said his current project would bring him to this facility several times a week for a while and was eager for details about OPP.

Ellen heard her name on the page and hurried to her office to find the light on the telset indicating a call had come into the Denver headquarters and had automatically been forwarded to her. She picked it up and found Cindy on the hot line with an emergency. They could not find the program for some revisions on a chip that was to be delivered to a customer within 48 hours. Cindy sounded stressed and very far away. Ellen felt the strain, too, but she had unswerving faith in the storage system for those chips. She would locate the backup and transmit it to Cindy within the hour. She told Cindy not to worry.

After notifying the Archives Department downtown that she would need the old auto files, she directed her attention to Cindy's problem. All the backup files were located in Denver, making it simple for Ellen to search them while the auto files were being scanned downtown. Her palms became a little clammy when the backup program she sought did not appear on the screen with the first try. She told herself she had entered an incorrect code somewhere and keyed in the instructions again.

Nothing.

With growing concern, she dialed a programmer in Denver and asked him to put the file on his screen. He couldn't access it either. He added another programmer to the call and all three of them began to search. The systems manager happened by somewhere along the line, and all four realized that the backup program simply was not there.

The systems manager began to stomp and rant about how they go to such great expense to protect the company's confidential transmission by encryption and then they lose something right there in the computer center. Ellen was vaguely aware that it may have fallen into the hands of a competitor, but no one wanted to face that possibility. One of Sims' top analysts had recently been convicted of computer crimes.

Cindy would be getting a little frantic by now. If there was anything Mr. Sims could not tolerate, it was late deliveries. Suddenly the whole idea of an orbiting factory became ridiculous—better to be close to home when something goes wrong.

Then Ellen remembered she had arranged a computer conference for several engineers from three western states on that very system. Could they provide any clues? She told everyone on the line she would check it out and call them back. As luck would have it, all three of the engineers were out of their offices—one on vacation, one at lunch, and one at an appointment with a customer. Ellen left voice messages in the electronic mailboxes of all three describing the problem, identifying the program by number, and stating that her call was extremely urgent. She called Cindy and told her to stay calm and then realized with a thud that it was way past lunch.

She had barely started the auto project, had not looked at the appliance cost lists, and was not sure whether the empty feeling in her stomach was because of the missing backup or because she was hungry. It was going to be a long day. She called home and reprogrammed the kitchen to prepare dinner for 7 instead of 6 o'clock and left a message on the recorder telling John and the kids to have some sandwiches if they got hungry.

Then she remembered that she had promised them she would teach them how to use the videotex library after dinner. Arthur, the oldest, had entered the fifth grade, and his teacher announced they would be required to use encyclopedias frequently for science class. They learned much more advanced technology now than had been taught when Ellen was in grade school. The biggest advantage to central library access was in the fact that encyclopedias were always current. Gone were the days when parents had to invest substantial sums in encyclopedia volumes that became obsolete in a year or so. Both Arthur and Susan liked interactive learning projects better than just reading from the screen, however, so Ellen knew they would require some supervision to maintain their attention. She made a mental note to call John after lunch and prepare him for the possibility that he might have to conduct the lesson alone.

A "ready" signal was blinking at her from downtown telling her the auto data she had requested was online and waiting for her. How that old machine annoyed her! She ignored it, logged out, and went to lunch, determined that she would at least get the addresses for the appliance cost system updated before she went home tonight.

She returned from her hasty lunch to find two messages displayed on the telset. One was from Mr. Sims, who undoubtedly wanted to know if she was making headway on the auto data, and the other was from the photocomposition machine operator, who was working on the current edition of the Sims Products Annual Report. She made a quick check of the auto data, saw no problems, crossed her fingers, and called Mr. Sims to tell him everything was on schedule. Then she called the typesetter, who reminded her that she had promised she would have the report edited before morning. Easy as it was to transmit the text to the pressroom by telephone lines, editing still required a lot of human concentration. This lost backup did not make it easy to concentrate, but she told the typesetter the copy would be in his hands on time.

Then she called Jim Tanaka, one of the senior proofreaders in the word processing center downtown, to see if he could handle the editing that afternoon. His line was busy, so she programmed her phone to signal her when it became free. At that point, she found it somewhat soothing to revert to the familiar old auto file and follow the old routines to produce the report that Mr. Sims had requested.

The warbling of the phone caught her off guard. Her heart leaped—it must be one of the engineers with good news about the backup.

Good news it was—but it was the signal saying Jim's line was free. Fortunately, her telephone had remembered the annual report—she had forgotten it. She dialed Jim's number, explained her situation, and successfully got the editing project under way. The burden of the annual report fell away, but she couldn't keep her eyes from traveling to the clock—3:30, and still not a word from the engineers. She didn't have the courage to face Cindy.

Finally, Dick Garcia, the engineer from Utah, called. He remembered the computer conference and he remembered the program. The customer was a real stickler, and they were having a problem with compatibility on one of the components. The engineer who had actually accessed the system was Meredith Butler, the one on vacation, and Dick wasn't sure how to locate her. He would see what he could do and call back. Ellen continued to plow through the auto file in an effort to control her nerves. Fully an hour later, the engineer called to say he had located Meredith at a resort on the Mediterranean and she was calling the computer now. She had a portable terminal with her and would probably be able to reconstruct enough of the conference to remember what was done with the backup. He would call Ellen back as soon as they had some developments.

Ellen couldn't leave Cindy hanging another minute. Three robots were idle without that program, and Cindy would probably want to reassign them if the delay was going to be much longer. Their phone conversation was brief, but inconclusive, and left them both still hanging in the air. Ellen went back to the auto file thinking she might have to cancel her dinner altogether.

Just as she was about to call John about the library project, the phone warbled again; one of the data searchers wanted to complain about some new software. Sarah always seemed resistant to change. Ellen didn't have a solution and found it difficult to listen. She tried to sound sympathetic and understanding, but finally had to terminate the call with a rather lame reference to the old auto data. Then, with a mounting headache, she decided to check her mail. Dialing the access code for her in-basket on the electronic mail system, she called the messages and documents up on the screen one by one. She made notes on some, printed hard copy of others, and simply read and deleted a few.

Actually, she wanted to print out two more, but she hesitated because she was so highly conscious of the current "Send It Without Paper" drive. The efforts in the late 20th century to reduce petroleum consumption were replaced by efforts in the early 21st century to reduce paper consumption. Paper was portrayed everywhere as an evil villain, destroying productivity and contributing to inflation. Still, people were slow to give it up. They were so accustomed to having something to touch and feel (or to touch and file!). Ellen was in no mood to make a decision, so she left those two documents in the system and called up her appointment calendar.

The next warble from the phone sounded genuinely hopeful for some reason. Ellen just knew something good had to happen soon. Sure enough, Meredith on the Mediterranean discovered she had inputted two digits incorrectly and filed the backup program in the wrong retrieval category. The error was easily corrected, the backup was ready for transmission to OPP, and the system in space would be up and running again within 30 minutes.

Two simple digits had sent Ellen on a merry chase all over the world and over 25,000 miles into space for the contents of a chip no bigger than a thumbnail.

"Two little digits!" she sputtered right out loud at the screen in front of her. The screen continued to blink its silent and relentless "ready."

Just then, Jim called with an easy question about the annual report. Ellen could tell he had it pretty well wrapped up, and she had very little left to do on the auto report. The appliance cost system could wait until tomorrow.

"How did you come out with your lost backup?" he inquired hesitantly.

"Fine," reported Ellen. "It was just a little input error, as usual. No problem." She could hear the relief in her voice as she spoke. Then she related the long day of telephone calls.

"You guys sure do get around out there!" was Jim's final comment.

Ellen hung up the phone with a startling realization of the critical role that the telephone had played in the little drama of the day. In the final analysis, however, she realized that the telephone, miracle that it represented, could not make up for the foibles of people. Even computers weren't foolproof. But, then, they weren't smart enough to get along without people either!

She picked up the phone and called John to say she would be home in 15 minutes.

■ EPILOGUE

This little story is not intended to be science fiction. Nor is it meant to be a prediction from a crystal ball. All the technology that is necessary to make it happen exists today. In fact, all of it is beginning to happen right now. The future is here, and if you are reading this book, you are part of it.

In the final analysis, it will enrich your life—if you are ready to learn and grow with it.

A Request for Your Opinion

There are no right or wrong answers to these questions. Compare your ideas with those of your classmates.

1. List the five most important contributions of telecommunications as we know it today.

Earth station. Photo courtesy of NEC America, Inc.

2. What do you think will be the most important telecommunications-related development of the future?
3. What do you think will be some of the hardships involved in living in a world of total communications?
4. Describe your idea of the role of knowledge workers in the offices of the future.
5. State how you feel developments in telecommunications might affect the career you have chosen for yourself.

Resources

■ RECOMMENDED BOOKS

ARREDONDO, LARRY A., *Telecommunications Management for Business and Government* (2nd ed.), New York: The Telecom Library, Inc., 1980.

ASTEN, KENNETH J., *Data Communications for Business Information Systems*. New York: Macmillan Company, 1973.

BROOKS, JOHN, *Telephone: The First Hundred Years*. New York: Harper & Row, Publishers, 1975–76.

CASSON, HERBERT N., *The History of the Telephone*. Chicago: A.C. McClury & Company, 1910.

Glossary of Terminology (PBX and Key Telephone Systems). Gaithersburg, Md.: The Marketing Program Service Group, Inc.

GRIESINGER, FRANK K., *How to Cut Costs and Improve Your Telephone, Telex, TWX and Other Telecommunications*, New York: McGraw-Hill Book Company, 1974.

KUEHN, RICHARD A., *Controlling Telephone Costs*. New York: AMA Management Briefings, 1972.

KUEHN, RICHARD A., *Interconnection: Why? How?* New York: AMA Management Briefings, 1975.

MARTIN, JAMES T., *Future Developments in Telecommunications* (2nd ed.), Englewood Cliffs, N.J.: Prentice-Hall, Inc., 1977.

MARTIN, JAMES T., *The Wired Society*. Englewood Cliffs, N.J.: Prentice-Hall, Inc., 1978.

MACMEAL, HARRY B., *The Story of Independent Telephony*. Chicago: Independent Pioneer Telephone Association, 1934.

SHERMAN, KENNETH, *Data Communications: A Users Guide*. Reston, Va.: Reston Publishing Company, Inc., 1981.

SHORT, JOHN, et. al., *The Social Psychology of Telecommunications*. London: John Wiley & Sons, Ltd., 1976.

SMITH, EMERSON C., *Glossary of Communications*. Chicago: Telephony Publishing Corporation, 1971.

TAYLOR, LESTOR D., *Telecommunications in Demand*. Cambridge, Mass.: Ballinger Publishing Co., 1980.

TOFFLER, ALVIN, *The Third Wave*. New York: Bantam Books, Inc., 1980.

TRAISTER, ROBERT J., *The Master Handbook of Telephones*. Blue Ridge Summit: TAB Books, Inc., 1981.

WINSBURY, REX, *Viewdata in Action*. London: McGraw-Hill Book Company (UK) Ltd., 1981.

■ RECOMMENDED PERIODICALS

Business Communications Review, 950 York Road, Hinsdale, IL 60521.
Communications Daily, 1836 Jefferson Place, N.W., Washington, D.C. 20036
Communications News, 124 South First Street, Geneva, IL 60134.
Interconnection, P.O. Box 1535, South Hackensack, NJ 07606.
Telecommunications, 610 Washington Street, Dedham, MA 02026.
Telephone News, 7315 Wisconsin Avenue, Ste. 1200 N, Bethesda, MD 20814.

■ ASSOCIATIONS

□ National and International Associations

International Organization of Women in Telecommunication (IOWIT)
 2554 Lincoln Boulevard #120, Marina del Rey, CA 90291
International Telecommunication Union
 Geneva, Switzerland
 (155 member countries)
Organization for the Protection and Advancement of Small Telephone Companies
 (OPASTCO)
 2301 M. Street N.W., Washington, D.C. 20037
National Telephone Cooperative Association (NTCA)
 2626 Pennsylvania Avenue, N.W., Washington, D.C. 20037
North American Telephone Association (NATA)
 1030 15th Street, N.W., Suite 360, Washington, D.C. 20005
 (interconnect industry association)

Society of Telecommunications Consultants
One Rockefeller Plaza, Suite 1410, New York, NY 10020
United States Independent Telephone Association (USITA)
1801 K Street, N.W., Suite 1201, Washington, D.C. 20006
United States Telecommunications Suppliers Association
33 North Michigan, Suite 1618, Chicago, IL 60601

□ Regional Associations

- American Hospital Association - Telecommunications Division
- Association of College and University Telecommunications Administrators
- Colorado Telecommunications Association
- Communications Managers Association of New York (CMA)
- Communications Systems Management Association
- Connecticut Telecommunications Association
- Georgia Telecommunications Association
- Greater San Diego Telecommunications Association
- International Communications Association (ICA)
- Michigan-Ohio Telecommunications Association
- Minnesota Communications Association
- Mississippi Valley Telecommunications Association
- New England Telecommunications Association
- North and South Carolina Telecommunications Association
- North Texas Telecommunications Association
- Professional Communications Management Association (PCMA)
- Southeastern Telecommunications Management Association
- Southwest Communications Association
- Tele-Communications Association (TCA)
- Telecommunications Association of the Northwest
- Telecommunications Management Association (TMA)
- Telecommunications Marketing/Sales Association
- Tennessee Telecommunications Users Association
- Wisconsin Telecommunications Association

■ REORGANIZATION OF AT&TS OPERATING COMPANIES

Pacific Telesis Group (Pacific Region)
Pacific Bell
Nevada Bell
US West, Inc. (Northwestern Region)
Northwestern Bell Telephone Company
The Mountain States Telephone & Telegraph Company
Pacific Northwest Bell Telephone Company
Southwestern Bell Telephone Company (Southwestern Region)

Ameritech Corporation (Great Lakes Region)
- Illinois Bell Telephone Company
- Indiana Bell Telephone Company, Inc.
- Michigan Bell Telephone Company
- The Ohio Bell Telephone Company
- Wisconsin Telephone Company

Bell South Corporation (Southeastern Region)
- South Central Bell Telephone Company
- Southern Bell Telephone & Telegraph Company

Bell Atlantic (Midatlantic Region)
- New Jersey Bell Telephone Company
- The Bell Telephone Company of Pennsylvania
- The Diamond State Telephone Company
- The Chesapeake & Potomac Telephone Company of Maryland
- The C & P Telephone Company of Virginia
- The C & P Telephone Company of West Virginia

NYNEX Corporation (Northeast Region)
- New York Telephone
- New England Telephone & Telegraph Company

■ THE LARGEST INDEPENDENT TELEPHONE COMPANIES IN 1980

- Anchorage Telephone Utilities, Anchorage, Alaska
- Ben Lomand Rural Telephone Co-op, Inc., McMinnville, Tennessee
- CP National Corporation, Concord, California
- Cameron Telephone Company, Sulphur, Louisiana
- Cencom, Inc., Rushford, Minnesota
- Centel Corp., Chicago, Illinois
- Century Telephone Enterprises, Inc., Monroe, Louisiana
- Chautauqua & Erie Telephone Corp., Westfield, New York
- Chillicothe Telephone Company, Chillicothe, Ohio
- Citizens Utilities Company, Stamford, Connecticut
- Clifton Forge-Waynesboro Telephone Company, Staunton, Virginia
- Commonwealth Telephone Company, Dallas, Pennsylvania
- Concord Telephone Company, Concord, North Carolina
- Conestoga Telephone & Telegraph Company, Birdsboro, Pennsylvania
- Conroe Telephone Company, Conroe, Texas
- Continental Telephone Corporation, Atlanta, Georgia
- Denver and Ephrata Tel. & Tel. Company, Ephrata, Pennsylvania
- East Assencion Telephone Company, Inc., Gonzales, Louisiana
- Eastex Telephone Cooperative, Inc., Henderson, Texas
- Ellensburg Telephone Company, Inc., Ellensburg, Washington
- Fairbanks Municipal Utilities System, Fairbanks, Alaska
- Farmers Telephone Cooperative, Inc., Kingstree, South Carolina
- Fort Bend Telephone Company, Rosenberg, Texas
- GTE Corporation, Stamford, Connecticut

- Golden West Telecommunications Co-op., Wall, South Dakota
- Grand River Mutual Telephone Corp., Princeton, Missouri
- Great Southwest Telephone Corporation, Grandview, Texas
- Guadalupe Valley Telephone Co-op, Inc., New Braunfels, Texas
- Guam Telephone Authority, Tamuning, Guam
- Gulf Telephone Company, Foley, Alabama
- Hargray Telephone Company, Inc., Hilton Head Island, South Carolina
- Heins Telephone Company, Sanford, North Carolina
- Horry Telephone Cooperative, Inc., Conway, South Carolina
- Illinois Consolidated Telephone Company, Mattoon, Illinois
- Kerrville Telephone Company, Kerrville, Texas
- La Fourche Telephone Company, Larose, Louisiana
- Lancaster Telephone Company, Lancaster, South Carolina
- Lexington Telephone Company, Lexington, North Carolina
- Lincoln Telephone & Telegraph Company, Lincoln, Nebraska
- Lufkin Telephone Exchange, Inc., Lufkin, Texas
- Mankato Citizens Telephone Company, Mankato, Minnesota
- Matanuska Telephone Association Inc., Palmer, Alaska
- Mid-Continent Telephone Corporation, Hudson, Ohio
- Mid-Plains Telephone, Inc., Middleton, Wisconsin
- Mid-Rivers Telephone Cooperative, Inc., Circle, Montana
- North Pittsburgh Telephone Company, Gibsonia, Pennsylvania
- North State Telephone Company, High Point, North Carolina
- Northern States Power Company, Minot, North Dakota
- North-West Telephone Company, Tomah, Wisconsin
- Pacific Telecom, Inc., Vancouver, Washington
- Pioneer Telephone Cooperative, Inc., Kingfisher, Oklahoma
- Puerto Rico Communications Authority, San Juan, Puerto Rico
- Puerto Rico Telephone Company, San Juan, Puerto Rico
- Rochester Telephone Corporation, Rochester, New York
- Rock Hill Telephone Company, Rock Hill, South Carolina
- Roseville, Telephone Company, Roseville, California
- San Marcos Telephone Company, Inc., San Marcos, Texas
- St. Joseph Telephone & Telegraph Co., Port St. Joe, Florida
- Sierra Telephone Company, Inc., Oakhurst, California
- Souris River Tel. Mutual Aid Co-op., Minot, North Dakota
- South Central Rural Tel. Co-op., Corp., Glasgow, Kentucky
- Standard Telephone Company, Cornelia, Georgia
- Sugar Land Telephone Company, Sugar Land, Texas
- Taconic Telephone Corporation, Chatham, New York
- Telephone and Data Systems, Inc., Chicago, Illinois
- Triangle Telephone Co-op. Association, Inc., Havre, Montana
- Twin Lakes Telephone Cooperative Corp., Gainesboro, Tennessee
- Universal Telephone Company, Inc., Milwaukee, Wisconsin
- Vista-United Telecommunications, Lake Buena Vista, Florida
- Virgin Islands Telephone Corporation, St. Thomas, Virgin Islands
- Warwick Valley Telephone Company, Warwick, New York
- Wood County Telephone Company, Wisconsin Rapids, Wisconsin
- Woodbury Telephone Company, Woodbury, Connecticut

Glossary

abandoned call — the caller gives up and hangs up after being connected and placed on hold for too long.

access code — the number or numbers dialed to connect to a certain circuit or service.

American standard code for information interchange (ASCII) — (pronounced "askie") a binary code recommended by the American Standards Association for use in teletypewriter exchanges and data communications.

amplifier — an active circuit designed to increase the power level of a given band of frequencies.

amplitude — the fullness or extent of a vibratory movement.

analog — transmission of a continuously variable signal.

area code — the three-digit codes that designate specific regions known as toll centers in the direct-dial, long-distance network (DDD).

attendant callback — the system signals the attendant when a call has been on hold for a specific number of seconds.

attenuation — power loss in a circuit due to distance.

authorization code — identification codes assigned to those persons authorized to use the telephone system, consisting of a digit or digits that must be dialed along with the number called.

automatic call distribution (ACD) equipment that automatically distributes incoming calls among specified stations in a specified order.

automatic dialing allows the user to store frequently called numbers and dial them automatically with a two- or three-digit code. Also called speed dialing.

automatic identification of outward dial calls (AIOD) an accounting system for monitoring outgoing toll calls automatically.

automatic private branch exchange (APBX) a private telephone switching system connected to the public network and capable of switching both internal and external calls mechanically.

automatic recall the user is signaled by electronic telephone equipment when a busy station becomes free.

band a range of frequencies.

bandpass The range of frequencies occupied by transmission.

bandwidth the range of frequencies available for a signal.

beeper a nickname for the radio transceiver used for remote paging that signals an incoming call with a beeping tone.

binary anything composed of two parts as in a binary code using only the two digits 0 and 1.

bit the smallest unit of data communications information, represented by on or off pulses in transmission.

bit stream a continuous stream of on-off pulses carried on a transmission line.

blocked call a call that encounters a busy signal.

boilerplating frequently used text stored in electronic media, such as a magnetic disk on a word processor, and printed automatically each time it is needed.

bps bits per second; a measure for determining the rate of transmission in data communications.

busy lamp field (BLF) a group of numbered lamps on a telephone set showing which stations on the system are in use and which stations are available to receive calls.

busy study monitoring of incoming telephone traffic to determine how many calls encounter busy signals; usually requires the telephone company's assistance.

cable (1) a group of twisted wires that carry communications signals; (2) a nickname for an international telegram.

call detail recording (CDR) a system for automatically recording information about toll calls, such as the station of origination, the circuit used, the time the call was placed, the length of the call, and the destination.

call director trade name for AT&T's key telephone set.

call diverter a device with which a user can have incoming calls automatically switched to another number.

call forwarding a feature that automatically switches calls to a designated number other than the number dialed.

call park a special hold mode that allows a caller to pick up a call at another station.

call pickup a feature allowing users to pick up their calls from stations on the system other than their own by dialing an access code.

call waiting a tone burst heard by a person on the phone, which announces that a call is waiting.

camping, or camp-on a hold mode that automatically rings through as soon as the station called becomes available.

carrier (1) a company that offers telecommunications transmission and related services; (2) a wire, cable, or radio beam that carries communications signals.

carrier current the electrical current that carries an intelligence signal along a transmission path.

carrier system a telecommunications path that can handle several transmissions simultaneously by a technique known as multiplexing.

cathode ray tube (CRT) the technical name for the visual display screen on electronic office equipment.

central office (CO) the carrier's facility where the customer lines terminate and where switching equipment is located.

CENTREX a private branch-exchange service provided by a telephone company, electronic switching equipment may be located on the telephone company's premises or on the customer's premises.

channel a path for electrical transmission for communications between two points.

chargeback accumulating costs for certain toll calls for the purpose of billing the customer, client or patient.

circuit a means of two-way communications between two points.

class of service the extent of toll service available to or authorized for certain stations or users on a system.

coaxial cable a carrier cable composed of as many as 22 tubes capable of handling many transmissions simultaneously.

common carrier a company that provides communications services to the general public and is subject to regulation as a public utility.

computer based message systems (CBMS) a system for inputting, delivering, and storing messages with communicating computers over telecommunications links; messages are displayed as images on screens.

computerized branch exchange (CBX) a private telephone switching system connected to the public network and capable of performing a wide variety of switching functions electronically.

conference call three or more parties on private or public networks (or a combination of the two) connected for one call; may require operator assistance.

consultation on hold a hold mode that allows a user to place one call on hold, make another call, and then return to the original call without redialing.

crossbar switching a mechanical switching system used for both public and private exchanges.

cryptor a device for converting data into a secret code during transmission.

customer dialed account recording (CDAR) an accounting system that charges customers or users for toll calls automatically by requiring an extra digit or digits to be dialed at the end of the calling number.

cycle one complete waveform.

data communications the systems and equipment for transmission of coded information over telecommunications channels.

decible (dB) a unit for measuring the loudness of a sound.

decoding translating information from coded form back to the original form readable by humans.

dedicated line or machine a machine or telephone line devoted exclusively to one user or purpose.

demodulating a technique which modifies a communications signal back to its original form (see modulating).

dial-up the use of a regular telephone line to connect to another station or terminal on a communications network.

diaphragm a vibrating disk used for transmitting and receiving telephone signals.

digital transmission transmission of discrete or discontinuous streams of on-off pulses.

direct distance dialing (DDD) a service that allows users to dial long-distance calls without operator assistance.

direct inward dialing (DID) a service allowing incoming calls to be dialed directly to an extension on the system without an attendant at a private branch exchange, such as PABX or CBX.

direct outward dialing (DOD) users on a private branch exchange can dial onto the public network without operator assistance.

direct station selection (DSS) allows an attendant to route calls by depressing a button for the correct station rather than using an intercom.

directory lookup system (DLS) a telephone directory stored in an electronic memory and accessed through a terminal with a CRT display; can also be linked to a printer to produce hard copy.

disks a form of electronic media.

distribution the way information is moved from one place to another.

do not disturb a feature on an electronic switch that allows the user to program his or her phone so neither incoming calls nor call waiting signals can get through.

dual tone multi-frequency (DTMF) the system of signaling the number desired in which each digit is represented by a tone composed of two frequencies, commonly known as Touch-Tone, which is AT&T's registered trade name.

dumb switch a mechanical switch.

echo check a method for checking the accuracy of transmissions used in data communications; data is returned to the sender for verification.

electronic computer-originated messages (ECOM) a satellite-based electronic mail service offered by the United States Post Office.

electronic document distribution (EDD) transmission of messages via telecommunication paths as electronic images rather than by mail; hard copy may be produced for the receiver. Also known as electronic mail.

electronic funds transfer (EFT) a method for transferring money over telephone lines without the use of paper.

electronic key telephone a telephone set used in conjunction with electronic or computerized telephone systems in which some functions are activated by pushing buttons or keys.

electronic mail See electronic document distribution.

electronic private branch exchange (EAPBX) a private telephone switching system connected to the public network and capable of switching both internal and external calls electronically.

elocution the art of effective speaking; generally denotes public speaking.

encoding converting a message from one language or system of communication to another.

encryption converting messages to unpublished or secret codes for transmission; commonly used in data communications.

ergonomics the science of fitting technology to people.

exchange a telephone switching center; the specific area served by a public exchange center.

executive override a programmable feature in electronic telephones that allows a certain station user to break into the calls of another.

extended area service (EAS) special telephone service offered by some common carriers in which local exchange rates are applied to calls made to nearby exchanges.

facsimile (FAX) a system for sending images over telephone paths; sometimes called telecopiers.

federal communications commission (FCC) the government agency that regulates the common carriers offering interstate communications services.

filter electronic circuitry which allows one part of a signal to pass while blocking another part.

flashing signaling a switch or exchange by depressing the switchhook on a telephone.

foreign exchange (FX) a trunk or line to an exchange office other than the local exchange office.

frequency the repetition rate of a periodically recurring waveform.

frequency division multiplexing (FDM) a technique for carrying more than one transmission on a line by dividing the frequency range into several bands.

full duplex transmission in both directions at the same time on the same line.

full screen incoming calls are controlled by an assistant so that nearly all calls are routed to persons delegated to handle specific areas of responsibility.

function key a button on a telephone set which, when depressed, causes the system to perform a task (such as creating a conference call or placing a call on hold).

giga a prefix meaning billion.

grade of service capacity of the subscriber service to receive calls; usually stated as a P factor. For example, a P factor of P01 would indicate that only one call in every 100 would encounter a busy signal.

half duplex transmission in both directions on one line but not at the same time.

hard copy information on paper.

helical wave guide a transmission carrier that is a metal tube in which very high frequency radio waves travel, capable of carrying 230,000 telephone conversations simultaneously.

hertz (Hz) a measure of frequency in cycles per second.

hold button a function key that places a call in a waiting mode.

hot line (1) a nickname for a tie line or private line, or (2) a service involving advice, information, or counseling over the telephone, usually without charge.

howler an alarm signal telling a user that the system has not accepted instructions or has been incorrectly operated in some manner.

hunting incoming calls search through specified station numbers until an available extension is found.

information the communication or reception of knowledge or intelligence.

information screen an attendant, assistant, or secretary intercepts incoming calls to obtain information about the call or caller before routing it to a station user.

integrated electronic office a combination of communicating electronic systems, equipment, and technologies that tie office operations together for efficient flow of information and work.

intelligence signal that portion of the electrical current in a transmission that carries information.

interactive system back-and-forth or inquiry-and-response communication between a user and a system.

interconnect industry firms other than telephone companies who develop, manufacture, and supply telephone equipment.

interconnection a connection between two telecommunications equipments and/or carriers.

international direct distance dialing (IDDD) an international service available in many countries allowing users to call foreign countries without operator assistance.

key telephone a multiline telephone on which the central office lines are accessed by depressing a button or key.

kilo a prefix meaning thousand.

laser a high-frequency light beam used in telecommunications applications for high-speed wideband transmission.

last number redial allows a station user to store the number dialed so it can be tried again automatically if the number is busy.

least-cost routing (LCR) capability for a system to automatically place a toll call on the cheapest available circuit; also called automatic route selection (ARS) or route optimization.

local call calls placed within the service area of a telephone company exchange.

local loop a channel connecting a subscriber or a subscriber's exchange to the central office.

local net a nickname for an intrafacility communications network.

measured charges the cost of a telephone call is based on the length of the call in minutes and the distance of the call in miles.

mega a prefix meaning million.

message unit (MU) an accounting method used by telephone companies for showing measured charges for local calls on a customer's bill.

message waiting a lamp or other signal telling a user to contact the attendant or operator for a message.

metacommunications implications in a statement not conveyed by verbal symbols.

microfilm a film on which very small images are stored.

microwave a radio carrier system.

microwave relay station a repeater station for microwave transmission carriers.

MODEM an acronym for a device that modulates and demodulates or converts a communications signal from one form to another.

modulating a technique that modifies the form of an electronic signal so that the signal can carry information on some form of communications media.

monitor the speaker on an electronic key telephone that allows the user to talk without lifting the receiver off the switchhook.

multiplex a technique for enabling a telephone path to carry several transmissions simultaneously.

music on hold a feature available for most phone systems that permits the caller to hear music while waiting on hold.

narrowband a communication channel with less than voice grade capacity used for low-speed data communication.

network control a system that enables management to control the quality and cost of the organization's telecommunications network and service.

night transfer a feature on an electronic switch allowing an attendant or user to program the system so that calls will ring into a different station during nonbusiness hours.

off-hook an activated telephone set; the receiver is off the switchhook.

off-premises extension (OPX) a tie line for connecting a location to the intercom of a private exchange located in another building.

on-hook a deactivated telephone set; the receiver is on the switchhook.

on-hook answerback allows a user to talk to an attendant without lifting the receiver off the switchhook.

on-hook dialing allows a user to leave the receiver on the switchhook while a call is being placed.

on-hook intercom allows a user to talk on the intercom line without lifting the receiver off the switchhook.

on-line two-way connection, usually via telephone links, with a computer.

other common carrier (OCC) another term for specialized common carrier; names the carriers competing against AT&T in long-distance telephone communications services.

override allows one user to break into the call of another.

P-factor a numerical representation of grade of service stating the percentage of calls that may be blocked.

panel switch a mechanical switching method in which a selector seeks the proper connection in a panel of terminals.

parity check a method for checking accuracy in data communications in which a mathematical solution is transmitted and verified.

partial screen incoming calls are controlled by an assistant so they can be accepted selectively and interruptions are minimized.

phonautograph a device that transcribes sound patterns into graphic form.

plain old telephone service (POTS) one phone with one line connected to a central office.

point of sale terminals (POS) payments by credit card are entered directly into the banking system.

point-to-multipoint a communications network that connects one location to several locations.

point-to-point a communication channel that connects only two points.

poller a device that automatically checks each station on a network to see if there are outgoing messages.

port an access point to a terminal or computer.

pounding the use of the # button to instruct a computerized phone system.

prefix that part of a telephone number identifying the exchange office.

private automatic branch exchange (PABX) see automatic private branch exchange.

private branch exchange (PBX) a private telephone switching system connected to the public network (see CBX, APBX, and EAPBX).

private line a telephone channel leased to a certain subscriber for exclusive use.

probe a telephone technique used to obtain information about a call or caller so the call can be routed correctly or handled more efficiently.

protocol the signals that determine format and instructions in data transmission.

public utilities commission (PUC) most frequently used name for the state agencies that regulate telephone companies and other utilities.

pulse code modulation (PCM) a technique for converting an analog signal to a digital signal by sampling the frequencies.

queuing holding incoming calls in line until a representative is available; holding outgoing calls in line until a low-cost circuit is available.

rate period a rate structure used by AT&T in which the cost of the call varies with the time the call is placed; lower rates apply during nonbusiness hours.

rate practices that part of the carrier's tariff publication containing the written statements and procedures on how rates will be applied.

rate schedules that part ot the carrier's tariff publication containing the rates to be charged for specific equipment or services.

record carrier companies that provide a communications service in which messages are transmitted by telecommunication paths with a document produced for the receiver.

record communication a communication service in which messages are sent by telecommunication path with a document produced for the receiver; telegraph, Telex, and TWX are examples.

remote call forwarding (RCF) allows callers from one city to dial a local number and reach a subscriber in a distant city; the call is charged at local rates while the subscriber pays a higher rate for the number.

repeater a device that amplifies or strengthens a communications signal that has weakened during transmission due to attenuation.

reprographics methods of duplicating hard copy output.

rotary turning on an axis like a wheel; the dialing method that sends a pulse to a switch for each number of positions a rotary dial is turned.

route optimization choosing the least-expensive route or circuit available at the time the call is placed.

signal the information or message transmitted by a communication system.

simplex transmission in one direction only.

soft page the attendant can page the user with a low-volume speaker on an electronic telephone.

speakerphone AT&T's registered trade name for its telephone set with a speaker allowing hands-free conversations.

specialized common carriers (SCC) carriers competing with AT&T in long-distance telephone transmission services.

speed dialing see automatic dialing.

starring using the * to instruct a computerized phone system.

station a subscriber's or user's location on a telephone system.

station callback a system that automatically notifies a station user when a call has been on hold a specified number of seconds.

station message detail recording (SMDR) systematic recording of usage details so that telephone costs can be accumulated according to the station placing the call, where the call went, and how long the call lasted.

step-by-step switch a mechanical switch in which a selector steps through an array of terminal connections as it receives pulses from a rotary dial phone.

switch a shortened name for the switching center where telephone traffic interconnects or transfers between circuits.

switchboard the board of telephone terminals from which a manual telephone exchange or switch is operated.

switched voice services long-distance carrier services competing against AT&T (see specilized common carriers).

switchhook the switch on the telephone set that is closed when the receiver is on the hook and opens the line when the receiver is off the hook.

tariff the published rate for equipment or service provided by a communications carrier and approved by the appropriate regulatory agencies.

telco a shortened name for telephone company.

telecommunications transmission of any intelligence signal by wire or radio or other electronic or electromagnetic carrier system.

telecommuting communicating with one's work place by telephone rather than by travel; work stations for tasks such as word processing or programming that may be placed in a remote location such as the employee's home.

telecopier photocopiers linked by telephone lines so the machine at one location accepts an image that is reproduced by the machine at another location; also known as facsimile.

telegram a message sent by telegraph.

telegraph a device or system for sending messages by electrical transmission of coded signals.

telephone tag a name for the problem of frequently incompleted calls due to unavailability of business people.

telephony converting speech sounds into electrical signals that can be carried over wires or radio beams.

teleprinter a teletypewriter.

teletext one-way videotex.

teletypewriter terminal equipment for keying input into a communications network.

telex a public teletypewriter exchange owned by Western Union.

telset short name for telephone set.

terminal equipment any device for sending and/or receiving information over a communications link.

tie line a private line provided by a carrier for the exclusive use of a subscriber.

time division multiplexing (TDM) a technique for combining several channels onto one line by interleaving the signals into time slots.

toll call a call made to a destination outside the local exchange.

toll center a telephone company central office where toll message circuits terminate.

toll restriction prevents stations from making certain long-distance calls.

trunk the circuits that connect switching centers.

TWX a public teletypewriter exchange.

United States independent telephone association (USITA) the organization for telephone companies that are not part of AT&T.

value-added carrier (VAC) another name for value-added network carriers.

value-added network carriers (VANs) specialized common carriers who offer telecommunication transmission services in conjunction with some time-shared computer services.

videoconferencing meetings of participants at remote locations via video and voice communication links.

videotex home televisions linked to remote computers using regular telephone lines.

viewdata another name for videotex.

visibility extent to which management can "see" the effective or ineffective operation of its communications network.

voiceband a telephone line suitable for transmission of speech; may be used for data, video, and FAX as well.

waveform complex sound waves created by speech or harmonic sounds.

wide area telephone service (WATS) a bulk-rate, long-distance service to specific areas either on an incoming line or an outgoing line (not both) offering cost advantages to high-volume users.

wideband a communication channel with a capacity that is greater than normal voice grade channels.

word/information processing the combination of people, procedures, and equipment that transforms ideas into readable communications and distributes them to final destinations.

WP an acronym for word processing.

zone charges measured charges for telephone calls to foreign exchanges.

zone paging allows a user to broadcast a message to a specific group of stations through an intercom.

zone unit measurement (ZUM) a billing system used by some telcos for calls to nearby exchanges.

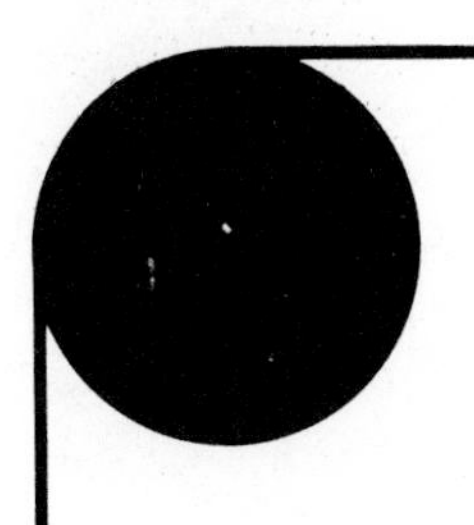

Index

D

E

U

V

W

Y

Z